# Macmillan Encyclopedia of the Environment

# Macmillan Encyclopedia of the
# *ENVIRONMENT*

VOLUME 6

*General Editor*
## Stephen R. Kellert

*Associate Editors*
## Matthew Black

## Richard Haley

*Macmillan Library Reference USA*
New York

Developed, Designed, and Produced by Book Builders Incorporated

Macmillan Library Reference
1633 Broadway, New York, NY 10019-6785

Library of Congress Catalog Card Number: 96-29045

Printed in the United States of America

Library of Congress Cataloging-in-Publication Data

Macmillan encyclopedia of the environment.
      p.       cm.
   "General editor, Stephen R. Kellert"—P. iii.
   Includes bibliographical references and index.
   Summary: Provides basic information about such topics as minerals, energy resources, pollution, soils and erosion, wildlife and extinction, agriculture, the ocean, wilderness, hazardous wastes, population, environmental laws, ecology, and evolution.
   ISBN 0-02-897381-X (set)
   1. Environmental sciences—Dictionaries, Juvenile.
[1. Environmental protection—Dictionaries. 2. Ecology—Dictionaries.]
I. Kellert, Stephen R.                96-29045
GE10.M33 1997                   CIP
333.7—dc20                      AC

Photo credits are gratefully acknowledged in a special listing in Volume 6, page 102.

# T

## Taiga

▶The northern CONIFEROUS FOREST, or FOREST of seed cone-bearing trees, of the Northern Hemisphere. Taigas are also known as boreal forests. The taiga is located in a band that runs across the Northern Hemisphere just below the ARCTIC Circle. Because of its location, the taiga is characterized by very cold temperatures and a winter season that lasts from six to ten months of the year. During this period, most PRECIPITATION falls as snow. Despite these harsh conditions, the taiga does have a short growing season that lasts for about two months. During this season, daylight is nearly constant.

### TAIGA PLANTS

The dominant plant life of the taiga are coniferous trees. The needles of these trees, their flexible branches, and their almost triangular shape help prevent heavy snow that could break limbs from collecting on the trees. In addition, a waxy coating on the leaf as well as a small surface area are ADAPTATIONS that help keep conifers from losing too much water. The shading provided by coniferous trees as well as acidic SOILS that form from DECOMPOSITION of their needles prevent most PLANTS from growing on the floor of the taiga. However, some mosses, FERNS, and small bushes, such as

◆ Moose are so common in some parts of the taiga that the region is sometimes called the "spruce-moose" belt.

blueberry, do survive here. In addition, LICHENS—organisms that result from a symbiotic relationship between FUNGI and ALGAE—commonly live on rocks and the trunks of conifers in the taiga.

### TAIGA ANIMALS

Some small lakes are scattered throughout the taiga. Swamps and bogs also become common during the growing season. These ENVIRONMENTS attract a variety of INSECTS and migrating BIRDS to the taiga. Several

larger animals, mostly MAMMALS, have adaptations that allow for living in the taiga year-round. For example, arctic hare are HERBIVORES that have a brown coat during the growing season. In winter, the hares display a white coat that acts as camouflage to protect them from PREDATORS such as lynx and gray wolves. Another large plant-eating mammal of the taiga is the moose. This animal has a very thick coat that insulates its body from the cold during the harsh winters. It also has large muscular legs and broad feet

◆ Debsconeag Lakes Forest in Maine is an example of a taiga.

adapted for MIGRATION through the snow. Despite its large size, the moose feeds largely on the tiny lichens it scrapes off rocks. Vegetation growing in lakes as well as on the ground is also used as food. [*See also* BIOME and DECIDUOUS FOREST.]

# Tailings

❚ **W**aste products, such as low-grade **ores** and heavy metals, that are removed as impurities from MINERALS extracted from Earth. Minerals such as COPPER and zinc are dug from the ground and separated from rocks and SOIL before they are used. This extraction and separa-tion process creates enormous amounts of waste. Often this waste is placed in piles, called *spoil piles,* near a mine. Part of the waste in a spoil pile consists of soil and rocks that had to be removed from the earth to reach the mineral deposit. The rest are tailings left after the metal has been separated from its ore.

Ores that are rich in minerals are called *high-grade ores.* Like many NATURAL RESOURCES, high-grade ores are in limited supply. As high-grade ores are used up, low-grade, less concentrated ores must be used to obtain minerals. Such ores create more tailings than higher-grade ores. A study conducted by the National Academy of Sciences pre-dicts that copper MINING in the year 2000 will produce three times as much waste for each ton of copper produced as copper mining did in 1978. In 1991, each ton of copper produced worldwide resulted in 110 tons (99 metric tons) of tailings from the copper ore. This study does not include the other waste matter, such as rocks and soil, con-tained in spoil piles that are removed from the earth just to reach the ore.

Spoil piles from mines are un-sightly. In addition, tailings con-tained in the spoil piles create many environmental hazards. For exam-ple, landslides created by such waterlogged mine wastes have destroyed entire communities.

Tailings and other mine wastes often pollute soil, air, and water. Wind carries particles of radioactive material, soot, dust, and toxic mate-rials in the air. Many of these mate-rials are harmful if breathed in or ingested (eaten) by living things. Pollutants that wash out or are leached from tailings can pollute soil and streams nearby. In the eastern United States, tailings from coal mines produce sulfuric acid. The acid is formed when water and air combine with iron sulfides in the mines. Rainwater falling on the spoil piles causes the acid to run off into streams. In the streams, the acid RUNOFF kills WILDLIFE and **corrodes** ships and bridges. In addition, the runoff dissolves heavy metals such as copper, zinc, arsenic, ALUMINUM, and magnesium from the tailings and the soil and carries them into waterways and groundwater. These heavy metals are toxic to organisms, making the water unfit for use.

Disposing of tailings without polluting the soil and water supply is a continuing problem. One method of disposal involves digging large lagoons, or artificial ponds, to catch water running off tailings. The water is treated with lime to make it less acidic. The process also causes toxic substances to settle to the bottom of the lagoon as **precipitates**. Period-

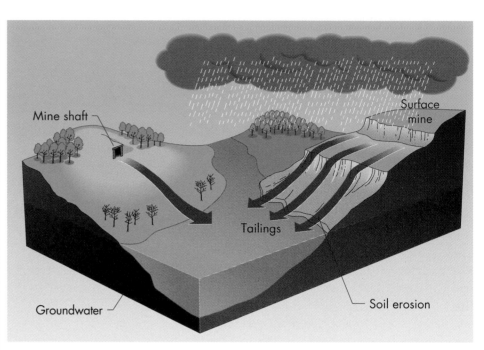

◆ Toxic runoff from tailings can pollute streams.

ically, the precipitates are removed when the lagoon is cleaned.

Another method used to prevent pollution caused by LEACHING is to plant vegetation on spoil piles. The vegetation helps to stabilize the pile and prevents runoff from car-

◆ Huge piles of tailings are left by a mine in Globe, Arizona.

rying toxic substances into nearby soil and streams. One difficulty with this method is that since mine waste is often toxic, PLANTS grow slowly or not at all. For instance, near Ducktown, Tennessee, a copper SMELTER operated until 1907. Nearly a century later, the surrounding soil is still so polluted that plants cannot grow in it.

In the United States, environmental laws regulate mining companies, which are required to control POLLUTION from tailings. However, in many other countries, there are no laws to prevent this type of pollution.

## Taylor Grazing Act

▶ Federal law that gives the government the responsibility for managing public GRAZING lands. The main purpose of the Taylor Grazing Act of 1934 is to protect PUBLIC LANDS from livestock OVERGRAZING. Overgrazing occurs when LIVESTOCK such as cattle, horses, and sheep are allowed to stay in one area for a long period of time and eat up the grasses and small PLANTS. When plants are plucked from the ground by feeding animals, or killed by animal trampling, SOILS loosen and lose their ability to hold together. Thus, overgrazing can increase soil EROSION and gradually transform a lush GRASSLAND into a bare DESERT.

◆ The federal government is responsible for managing public grazing lands under the authority of the Taylor Grazing Act of 1934.

Under the act, the Federal Grazing Service was established in 1939 to develop strategies for protecting public lands from this kind of damage. Today, the BUREAU OF LAND MANAGEMENT and the U.S. FOREST SERVICE manage all public grazing lands in the United States. [*See also* DEPARTMENT OF THE INTERIOR; DESERTIFICATION; DUST BOWL; HERBIVORE; MULTIPLE USE; and NATIONAL GRASSLAND.]

## Telkes, Maria (1900– )

▶ Hungarian-American physical chemist who was a pioneer in the use of SOLAR HEATING in buildings. Maria Telkes was born and educated in Budapest, Hungary. As a high school student, Telkes developed an interest in the possible use of SOLAR ENERGY as a power source. Her interest led her to pursue a career as a physicist. In 1924, Telkes received her doctoral degree in physical chemistry. The following year, she came to the United States to visit an uncle, and she remained in the United States for the duration of her career.

From 1926 until 1937, Maria Telkes worked as a physicist with the Cleveland Clinic Foundation. While there, Telkes helped to develop a PHOTOVOLTAIC CELL that was used to record the electrical energy given off by the human brain. The same year that she left the foundation, Telkes became an American citizen.

During World War II, Maria Telkes began working for the government of the United States as an adviser to the Office of Scientific Research and Development. While working in this position, Telkes developed a method for using solar energy to remove the salts from sea water to make the water fit for drinking. The device was installed in life rafts for use by soldiers fighting in the war.

During the 1940s, Maria Telkes began focusing her attention on how to use solar energy to heat buildings. Her work resulted in the construction of one of the first solar heated homes in Dover, Massachusetts. The system, which made use of chemical substances and blowers to control heat absorption and circulate air, was one of the earliest and most successful examples of an active solar heating system. [*See also* ALTERNATIVE ENERGY SOURCES and WATER, DRINKING.]

# Tennessee Valley Authority (TVA)

▶ An agency of the U.S. government designed to promote development in the Tennessee River Valley and nearby areas. The TVA was created in 1933 as part of the New Deal of President Franklin D. ROOSEVELT's administration. Its task was to develop HYDROELECTRIC POWER in northern Alabama and to provide navigation and flood control on the Tennessee River. The TVA is administered by a three-member board appointed by the president of the United States with the Senate's consent.

The TVA is best known as a provider of ELECTRICITY. Although it was formed to develop hydroelectric power, it later built coal-fired and nuclear power plants as well.

There are 50 DAMS under the authority of the TVA. When it built the Tellico Dam, it ran into opposition from people concerned about the snail darter. The snail darter is a tiny SPECIES of FISH that was considered endangered. The dam's completion was delayed for several years due to debates on how the

◆ In 1933, the TVA was created to develop hydroelectric power and to supply flood control on the Tennessee River and in nearby areas.

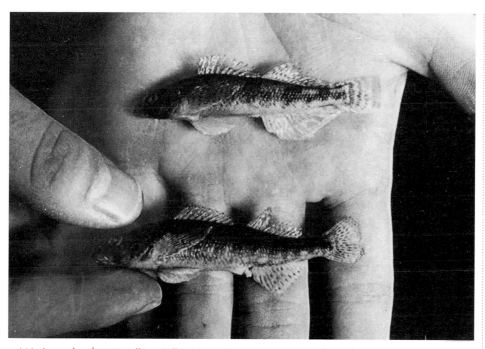

◆ Work on the $115 million Tellico Dam in Tennessee was halted when it was discovered that its completion could lead to the extinction of the snail darter.

unfair competition. It was once hoped that the TVA might prove a model for progress in underdeveloped parts of the country, but this now seems unlikely.

## Terracing

▮ A soil CONSERVATION method in which crops are grown on a series of broad, steplike terraces carved into a slope. In areas that have steep, hilly terrains, terracing is a common farming practice. SOILS of such terrains are more prone to EROSION than are soils located on flat terrains. On sloped terrains, the

ENDANGERED SPECIES ACT should be applied to this situation. Although the dam was built, the snail darter did not become extinct.

While the TVA has faced opposition, it has also brought about many positive accomplishments. It has aided farmers by finding ways to improve farm yields and reduce soil EROSION while developing new fertilizers. The TVA has also been active in the conservation movement. It has been responsible for the reforestation of more than 1 million acres (400,000 million hectares) of land in the valley. The TVA has brought additional revenue to the region by developing recreational areas near the lakes and RESERVOIRS it created.

The TVA has always had critics, such as private utility companies, which complain that it provides

◆ Terracing is an effective soil conservation technique. By trapping water as it runs down steep slopes, terracing is also a useful method for growing water-loving plants, such as rice.

force of gravity allows water to run quickly downhill, permitting rapid erosion. Fast-moving water can loosen and carry away soil. Thus, terracing is extremely common in parts of the world that experience heavy rains, such as South America, China, Japan, and the Philippines.

Terracing is effective at reducing soil erosion because the nearly level terraces slow water as it runs down a slope. As the water's energy of motion is reduced, its ability to erode soil is also reduced. When an area makes use of terracing, water flows gradually down a slope from one terrace to the next.

Terracing is also a good technique for growing crops that require a great deal of moisture, such as rice. When water runs down a terraced hillside, it gets trapped on each terrace for a period of time. Any water that runs off one terrace cascades to the next terrace. In this way, large pools of water are created on the hillside to provide an ENVIRONMENT that is perfect for growing rice and other plants that require high moisture. [*See also* CROP ROTATION; NO-TILL AGRICULTURE; RUNOFF; and SUSTAINABLE AGRICULTURE.]

# Thermal Water Pollution

I▶ POLLUTION caused by a large increase in water temperature due to human activity. Power plants,

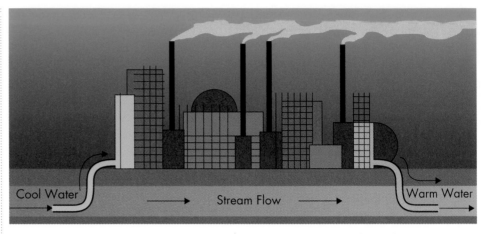

◆ The discharge of heated water into a river, lake, or stream is a form of pollution.

petroleum refineries, paper mills, and other industrial facilities often use water cooling systems to reduce excess heat.

Cool water from lakes, rivers, or bays is pumped into pipes that lie beside pipes containing hot water from a plant or factory. Heat is transferred from the hot water to the cool water. Once cooled, the water is returned to the plant. The cool water, now heated, is sometimes released back into the original source.

Another source of thermal water pollution may be organic products that give off heat as they decompose. If a great deal of DECOMPOSITION is occurring in an area near a waterway, such as a LANDFILL, enough heat can be produced to raise the temperature of the water to a point where it can harm organisms.

Thermal water pollution can adversely affect water quality and aquatic life. An increase in water temperature decreases the amount of DISSOLVED OXYGEN the water can hold. FISH suffocate because they cannot get enough OXYGEN. The

increased water temperature is also destructive to young fish and developing eggs.

# Thermodynamics, Laws of

I▶ Fundamental laws of nature that help scientists describe how matter and energy flow through the natural world. The laws of thermodynamics were first described by nineteenth-century scientists studying heat engines. Later, it was discovered that the laws apply to all forms of energy. The laws of thermodynamics have helped scientists understand how ECOSYSTEMS work, how PLANTS capture the energy in sunlight, and how energy captured by PLANTS becomes available to other organisms. These fundamental laws have also enabled scientists to develop new technologies that make use of the "wasted" energy

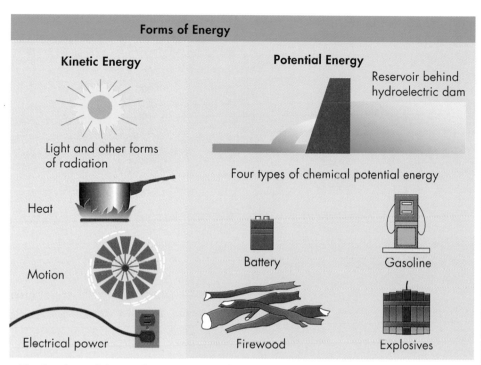

**Forms of Energy**

**Kinetic Energy**

Light and other forms of radiation

Heat

Motion

Electrical power

**Potential Energy**

Reservoir behind hydroelectric dam

Four types of chemical potential energy

Battery

Gasoline

Firewood

Explosives

◆ The first law of thermodynamics states that energy can be converted into its various forms.

produced by power plants as energy is changed from one form to another.

## FIRST LAW OF THERMODYNAMICS

The first law of thermodynamics states that energy can be changed from one form to another but is neither created nor destroyed in the process. Energy comes in many forms. These forms include: thermal (heat) energy, electrical energy, light energy, nuclear energy, chemical energy, and mechanical energy. To illustrate how energy is changed from one form to another, consider a light bulb. When a switch is turned on, ELECTRICITY begins running through the bulb. When the electricity reaches the bulb filament—the thin metal wire inside

the bulb—the electricity changes to heat energy. Some of the heat energy is changed to light energy as the filament begins to glow. In this instance, electrical energy is converted into both heat and light energy.

Another familiar example of energy conversion occurs during the striking of a match. When a person strikes a match, he or she uses mechanical energy to rub the match against a surface. Heat energy, produced by the friction between the match and the surface, causes a chemical reaction in the match head. The chemical reaction causes the match to ignite and burn. Heat and light energy are contained in the flame.

Scientists use the first law of thermodynamics and the fact that energy can be changed or con-

verted into different forms to explain the ecological concept of the FOOD CHAIN. In the food chain, energy from the sun is changed by plants and other PRODUCERS into the chemical energy of food. Through this process, known as PHOTOSYNTHESIS, producers make this chemical energy available to all other organisms in the ecosystem. The chemical energy is passed through the food chain as organisms eat one another.

The first law of thermodynamics has other useful applications, as well. For example, by using this law, one can calculate the amount of heat energy given off when a certain amount of COAL is burned. One could then calculate how much electrical energy could be generated by the heat. However, such calculations would be impossible to make without considering the second law of thermodynamics.

## SECOND LAW OF THERMODYNAMICS

The second law of thermodynamics states that when energy is converted from one form to another, some amount of energy is always "lost" or wasted. This law is familiar to anyone who has ever participated in strenuous physical activity. For example, when a person exercises, the chemical energy stored from the foods he or she ate is converted into the mechanical energy of movement. Because the conversion is not completely efficient, some energy is lost as heat—which is why people get warm when they exercise.

The second law of thermodynamics helps ecologists understand how energy is used by the organ-

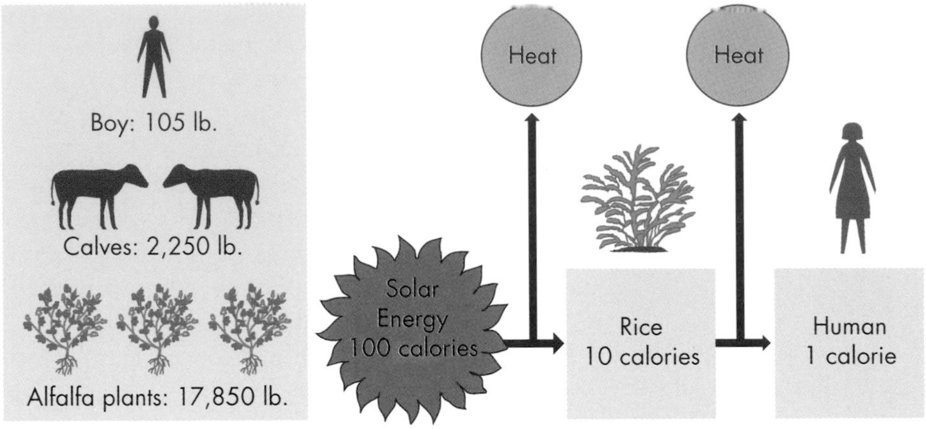


◆ According to the second law of thermodynamics, when energy is converted from one form to another, some energy is lost as reflected in the food chain above.

isms in an ecosystem. Through a model known as an ENERGY PYRAMID, scientists show that in a food chain the amount of energy available to each succeeding TROPHIC LEVEL decreases. In fact, the total energy transfer from one trophic level to the next is only about 10%. According to the second law of thermodynamics, this energy loss results because some energy is used by an organism or lost to the ENVIRONMENT as heat before it is passed to the next trophic level. The energy pyramid also shows that the number of organisms at each trophic level in an ecosystem decreases at succeeding trophic levels. Thus, there are fewer organisms feeding at the top of the food chain than at lower levels of the food chain. In this way, the second law of thermodynamics places limits on the number and types of organisms that can make up an ecosystem.

Scientists also use the second law of thermodynamics to develop methods for improving ENERGY EFFICIENCY. For instance, in nuclear power plants, only about 30% of the energy available in nuclear fuels is used to produce electricity. The remaining 70% is lost to the environment as heat. Similarly, large amounts of heat are lost when coal and other FOSSIL FUELS are burned to generate electricity. Scientists are now developing ways to capture this wasted heat energy so that it can be used to heat homes and businesses. [*See also* ALTERNATIVE ENERGY SOURCES; COGENERATION; and ECOLOGY.]

# Thoreau, Henry David (1817–1862)

American writer and naturalist, born in Concord, Massachusetts, from which he seldom strayed far. After graduating from Harvard, Thoreau briefly taught school, was Ralph Waldo Emerson's houseguest for a time (1841-43), and, between July 4, 1845, and September 6, 1847, lived and wrote in a 10-by-15-foot (3-by-4.5-meter) hut he had built by Walden Pond, at Concord. His account of this 26-month period, *Walden, or Life in the Woods,* was published in 1854 and has remained popular to this day. Another of his works, an essay entitled "Civil Disobedience," helped to inspire the nonviolent campaign for India's independence during the twentieth century. Similarly, Thoreau's impassioned advocacy of individualism, the simple life, love of nature, and social protest continues to inspire modern environmental activists throughout the world.

# Three Mile Island

The site of a nuclear power plant along the Susquehanna River near Middletown, Pennsylvania, where the worst nuclear accident in the United States took place. On March 28, 1979, Three Mile Island's nuclear reactor began to discharge "puffs" of **radioactive** gas into the ATMOSPHERE. The accidental emissions occurred as a result of human and mechanical errors. A malfunction of the plant's cooling system was worsened both by problems with computer monitors and human supervision of the system. These problems resulted in a breakdown of the reactor's cooling system and the destruction of the reactor core—the center that contained the radioactive material.

Early news bulletins, with statements from plant management and local officials, minimized the seriousness of the problem. However, it was soon discovered that a potentially explosive hydrogen bubble had formed in the reactor. The formation of the bubble made a **core meltdown** possible. It was also discovered that neither the plant nor city officials had created a workable plan for evacuating people from areas surrounding the plant in the event of an accident. In all, 200,000 people had to be evacuated from the area to get them out of harm's way.

Scientists and technicians worked for 12 days to contain the damage at the Three Mile Island plant and to prevent a total core meltdown. Such a meltdown might have permitted huge amounts of radioactive material to enter the ENVIRONMENT. Because RADIATION is harmful to living things, a meltdown threatened people and WILDLIFE.

The Three Mile Island meltdown was prevented. No fatalities or known injuries resulted from the incident. However, cleanup of the contaminated area in and around Three Mile Island continued through 1989. The cleanup efforts, along with the potential danger of radiation in the environment, justified the concerns of environmentalists who warned about constant threats of radiation from other nuclear plants.

A far worse nuclear incident occurred in 1986 when the world's worst nuclear power plant disaster

took place at Chernobyl in what was then the Soviet Union. An explosion tore apart the plant's reactor. Huge amounts of radiation were discharged into the atmosphere, immediately killing more than 30 people and seriously injuring 200 others. Some doctors estimate that more than 800,000 children who lived in that area are at risk for developing CANCER. These children inhaled radioactive iodine or drank contaminated milk before they were evacuated from

the area. Radioactive material from Chernobyl, carried by the wind, fell on parts of northern and central Europe. Millions of people still live in contaminated areas around Chernobyl because their governments do not have money to relocate so many people.

Nuclear plants routinely release some radiation into the air. However, the amounts are generally too small to be harmful to living things. The Three Mile Island disaster made Americans more aware of the potential dangers at plants using radioactive material. Some environmental groups want the use of nuclear energy stopped. Since the Three Mile Island incident, no new nuclear reactor construction has been started in the United States. [*See also* ATOMIC ENERGY COMMISSION; CHERNOBYL ACCIDENT; HAZARDOUS WASTE; NUCLEAR POWER; and RADIOACTIVE WASTE.]

# Tidal Energy

◗ Energy produced by the periodic rise and fall of ocean waters caused by the gravitational forces of the moon, sun, and Earth. Along most seashores, tides slowly rise and fall a few feet twice a day. The rise in water levels is known as high tide. The fall in water levels is known as low tide. Tidal power plants harness the energy from the constant rise and fall of tides to produce ELECTRICITY.

 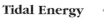

◆ Tidal power plants like this one use the tremendous energy of the moving tides to generate electricity.

## HOW A TIDAL POWER PLANT WORKS

A tidal power plant is an example of a HYDROELECTRIC POWER plant. In this type of plant, electrical energy is generated when water flows over a turbine. A turbine is similar to a fan with rotating blades. When water pushes the blades, the turbine spins. This spinning action is then used to turn a generator to produce electricity. In a tidal power plant, the in and out flow of the tides spins the turbine generator to produce electricity. Power plants using tidal energy already exist in Canada, China, and Europe.

Tidal energy could become an important source of energy in the future because it is an example of a RENEWABLE RESOURCE; that is, it will never run out. In addition, tidal energy, like SOLAR ENERGY and WIND POWER, is a relatively "clean" source that does little damage to the ENVIRONMENT. [*See also* ALTERNATIVE ENERGY SOURCES; FOSSIL FUELS; NATURAL RESOURCES; and OCEAN.]

# Tiger

◗A CARNIVORE that is the largest of the great cats. It lives in the FORESTS, marshes, swamps, and GRASSLANDS of Southeast Asia and faces EXTINCTION due to overhunting and loss of its natural HABITAT. The tiger *(Panthera tigris)* is a good swimmer and on hot days spends much time in the water. It can climb trees, but rarely does. It has a roar that can be heard 2 miles (about 3 kilometers) away. Unlike the lion, which roams in open country, a tiger usually stays in the shadows, where its - protective coloration helps to camouflage it among tall grass and trees. A tiger's coat ranges from red-orange to brownish-yellow, with black markings, and whitish fur on the throat, belly, and insides of the legs.

The average adult male tiger is 9-feet (2.7-meters) long, including its 3-foot (0.9-meter) tail, and weighs about 420 pounds (190 kilograms). A female averages 8 feet (2.4 meters) in length and weighs about 300 pounds (135 kilograms). The female tiger produces a litter of two or three cubs that stay with her for about two years, until they can successfully hunt food for themselves. This carnivore usually preys on large animals, such as deer, antelope, and wild oxen, but also eats smaller prey, like monkeys and frogs. The tiger often preys on porcupines, even though their quills may cause the PREDATOR painful wounds. Sometimes a big cat will kill domestic cattle, if hunters have reduced the number of wild prey available. A human-eating tiger is rare—possibly 3 out of every 1,000 tigers. It is usually a wounded animal that can no longer hunt faster prey, or one forced to find any food it can when there is not enough available game. Where tiger preserves and dense human populations are close to each other, however, tiger attacks are more common.

The tiger is a night hunter that uses its sharp eyes, keen ears, and good sense of smell to find prey. The tiger conceals itself until the prey is close by, then leaps out to attack it. The powerful predator drags its prey to thick cover, often hauling a 500-pound (225-kilogram) animal up to 3 miles (4.8 kilometers). There the tiger eats all of its prey, except the bones and stomach, and may consume up to 50 pounds (23 kilograms) in a single night.

## TIGERS IN DANGER

In the early 1900s, there were more than 100,000 wild tigers in Asia; today there are fewer than 5,000. Tigers lose their habitat to humans, who strip the forest of timber and MINERALS and leave leveled land, onto which farmers move. Each year, many of the big cats are killed by poachers—illegal hunters—who sell tiger parts for big money. In China and Taiwan, powdered tiger bone that is used in medicine sells for $500 a gram, and other tiger parts may sell for as much as $1,700.

The CONVENTION ON INTERNATIONAL TRADE IN ENDANGERED SPECIES OF WILD FAUNA AND FLORA (CITES) of 1989 banned the trade in tiger products, but the ban is difficult to enforce in countries where tiger parts have been consumed for centuries. If the decline continues, experts say the wild tiger could be extinct before the year 2000. [*See also* BIODIVERSITY; ENDANGERED SPECIES; and HABITAT LOSS.]

◆ The habitat of the Bengal tiger stretches from India across Southeast Asia.

# Topsoil

▐▶ The loose uppermost layer of SOIL in which the roots of most PLANTS grow. Topsoil is composed of rock particles, air, water, and decayed organic matter. Many organisms, such as worms, INSECTS, FUNGI, and BACTERIA, also live in topsoil. Most of these organisms help make the soil more suited to the growth of plants by creating spaces between soil particles through which air and water can move or by breaking down organic wastes through the process of DECOMPOSITION. The actions of such organisms aid in making the soil **fertile** and **arable**.

Because most plants grow in the topsoil layer, this layer of soil is vital to farmers. However, much of Earth's topsoil is being affected by damaging natural processes and human activities. One process that greatly affects topsoil is EROSION, or the carrying away of soil by the wind, water, or other agents. Erosion is a natural process. However, some activities of people increase the likelihood that valuable topsoil will be lost to erosion. Such prac-

certain types of nutrients from soil. [*See also* CONTOUR FARMING; NO-TILL AGRICULTURE; OVERGRAZING; and SOIL CONSERVATION.]

**arable** capable of supporting the growth of crops.

**fertile** having the texture, nutrients, and other qualities that allow for abundant plant growth.

tices include the plowing of land by farmers, which removes vegetation from and loosens soil, making the soil lighter and more easily carried away by wind or running water. The raising of grazing LIVESTOCK that are allowed to remove too much vegetation from soil also helps speed the action of erosion.

Many farming practices also harm topsoil by removing too many nutrients from the land. For example, when the same crops are grown on a piece of land year after year, the same nutrients will continuously be removed from its topsoil. Over time, as the nutrient levels decrease, the soil may become unable to support the growth of some types of crops. Many farmers use synthetic fertilizers to prevent this problem. However, the use of such fertilizers often leads to other environmental problems. Additional farming methods, such as CROP ROTATION, help prevent the loss of too much of one type of nutrient from the land. In this farming method, the parcels of land on which different types of crops are planted are rotated from year to year. In this way, crops that have the same nutritional needs are not planted on the same land each year, which decreases the loss of

# Toxic Substances Control Act (1976)

▶ A federal law that gives the U.S. ENVIRONMENTAL PROTECTION AGENCY (EPA) the job of controlling the production, use, and disposal of chemicals that may be hazardous and are not controlled by other federal laws. There are federal laws to control the uses of special classes of chemicals used in drugs, foods, cosmetics, PESTICIDES, tobacco, and other products. Taken all together, though, these laws do not control every kind of chemical that is produced by industries or laboratories. The Toxic Substances Control Act of 1976 (also known as TSCA or TOSCA) includes chemical products that are new; chemicals that are not new, but will be put to a new use; chemicals that are newly imported from other countries; and chemicals not already controlled by other laws when the TSCA was passed.

In the case of a new chemical product, the TSCA requires that an industry give information to the EPA about any known effects of the chemical on human health and

◆ People working with toxic wastes must protect their lungs and the outside of their bodies from contact with the material.

the ENVIRONMENT. This must be done before the chemical is made available to the public. The industry must also keep records of how the chemical product affects the workers who handle it. The EPA must review this information and decide whether the chemical is a risk to human health or to the environment. If it seems to be a hazard, the EPA has the authority to control the way the chemical is produced, used, or disposed of. If the chemical is judged to be an "unreasonable risk," the EPA can also require testing of the chemical to find out if it causes CANCER or other ill effects.

A chemical that has already been in use for a long time can still be controlled under the act. For example, polychlorinated biphenyls (PCBS) were in use long before the act was passed. However, these chemicals proved to be hazardous, and their use is not covered by other laws. The TSCA gives the EPA the authority to control the disposal of PCBs.

The job given to the EPA by the TSCA is a very big one. Because of TSCA, the EPA is expected to keep track of the risks of over 65,000 chemicals that are in use. It must require whatever tests or controls are needed to protect people and the environment from these substances. It is also required to do research on the effects of some chemicals on humans. In addition, it works with other government agencies, such as the Occupational Safety and Health Administration, that share EPA's job of controlling hazardous chemicals. [*See also* COMPREHENSIVE ENVIRONMENTAL RESPONSE, COMPENSATION, AND LIA-

BILITY ACT (CERCLA); HAZARDOUS SUBSTANCES ACT; LAW, ENVIRONMENTAL; and RESOURCE CONSERVATION AND RECOVERY ACT (RCRA).]

## Toxic Waste

▶ Any liquid, solid, or gas that is harmful or fatal when swallowed, digested, absorbed, or inhaled. Some examples are ASBESTOS, arsenic, MERCURY, and LEAD.

The damaging consequences of toxic waste do not always show up right away. Some may not show up for a few hours and others may not show up for years, but whenever they do, they threaten Earth's ecological balance. Toxic waste disposed of on land, or buried under it, may leach from the material, polluting not only the land but groundwater as well.

About 50% of the U.S. population gets its drinking water from groundwater, not SURFACE WATER. Groundwater sources form from rain and snow that fall over many years, filtering down through the SOIL and rocks to collect underground. The water settles in vast stretches of absorbent rock, sand, and gravel, called AQUIFERS. The stored water is basically pure—

◆ Before the EPA's crackdown on polluters, manufacturers stored toxic waste in any convenient spot, such as the back of a trailer or an abandoned mine shaft, where leaks from the toxic contents poisoned rivers and underground water sources.

largely bacteria-free and chemical-free. In the United States, there is roughly five times as much water in such underground holding ponds as flows through all the country's surface rivers, lakes, and streams in one year.

## TOXIC CHEMICAL WASTE

Toxic chemicals, both natural and manufactured, are harmful to living organisms. Some harm only certain PLANTS or animals but not humans; others harm only humans. Many kinds of human activity, such as farming, MINING, energy production, manufacturing, and even running a household, require the making or using of toxic chemicals. Homes and businesses use everything from drain openers and paint thinners to PESTICIDES and rust solvents.

The Toxic Release Inventory (TRI), established by the 1987 amendments to SUPERFUND, monitors

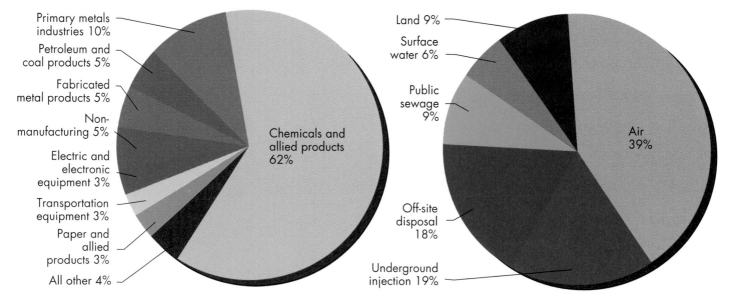

THE LANGUAGE OF THE ENVIRONMENT

**injection wells** shafts encased in steel and concrete deep underground into which toxic materials are inserted by force and pressure.

the release of toxic chemicals into the ENVIRONMENT by manufacturers with more than ten employees. In 1989, TRI reported that 22,650 manufacturers released a total of 5.7 billion pounds (2.6 billion kilograms) of toxic chemicals: 2.4 billion pounds (1.1 billion kilograms) were emitted into the air; 1.2 billion pounds (.5 billion kilograms) were buried in **injection wells**; 189 million pounds (86 million kilograms) were released into streams, lakes, and rivers; and 445 million pounds

(202 million kilograms) were buried in LANDFILLS.

The RESOURCE CONSERVATION AND RECOVERY ACT (RCRA) of 1976 and later amendments regulate large companies that produce toxic waste materials. The companies must report which toxic materials they use, where toxic wastes will be taken for disposal, and by what method disposal will be carried out. In this way, RCRA can track the production and disposal of toxic waste "from cradle to grave."

RCRA does not cover companies producing less than 2,200 pounds (990 kilograms) of toxic waste a month, such as gas stations, dry cleaners, food establishments, and other small businesses. Environmentalists believe that these exempted companies together add another 1.3 billion pounds (0.6 billion kilograms) of toxic chemical wastes into the air, water, and soil each year.

◆ The chart on the left shows the percentage of toxic wastes produced by the various types of industries. The chart on the right shows the percentage of toxic chemicals emitted into the air, land, and water by manufacturers.

After years of dumping toxic chemical waste, people discovered the hard way how such deadly materials can affect groundwater. During the 1940s and 1950s, thousands of tons of toxic chemical wastes were buried in a landfill in the LOVE CANAL area of Niagara Falls, New York. Years later, after a housing development and elementary school had been built on the old landfill, toxic chemicals oozed out of the ground. Residents suffered unusually high incidents of CANCER, miscarriages, and birth defects. In 1978, it was discovered that toxic waste leaching from the dump had contaminated the area's groundwater. Love Canal's residents were evacuated, and the area was condemned as unfit for human existence.

◆ The EPA is cleaning up toxic wastes in Denver, Colorado. Investigating agents are present to oversee the cleanup.

## DISPOSING OF TOXIC WASTE

Most toxic waste is disposed of either on or near the site where it originated. However, U.S. companies export more than 160,000 tons (144,000 metric tons) of such waste each year to Canada, Mexico, or to very poor developing nations, which have no safe-disposal rules but do have a desperate need for money to keep their countries running. In March of 1989, the United States signed the United Nations treaty on the Control of Transboundary Movements of Hazardous Wastes and Their Disposal, designed to outlaw international trade in toxic waste. However, in its final form the treaty seemed to authorize trade in some toxic wastes, including RADIOACTIVE WASTE, as long as the nation to which it was sent agreed to accept it.

## SOURCE REDUCTION

Minimizing the production of toxic waste became a federal policy with the 1984 amendments to RCRA and the setting up of the ENVIRONMENTAL PROTECTION AGENCY (EPA) Office of Waste Minimization in 1987. In response, the Council of State Governments (CSG), a national organization representing the 50 states, developed examples of different types of state laws that could be put in place to reduce HAZARDOUS WASTE. Many states enacted such laws, including a program in North Carolina called "Pollution Prevention Pays" that gives technical advice, research information, and matching grants to cities and industries that want to achieve waste reduction and recycling.

In 1991, the EPA started a voluntary program called the Industrial Toxics Project (ITP). It asked businesses to cut emissions of 17 chemicals listed on the Toxic Release Inventory by 50% before the end of 1995. To kick off the program, the EPA sent letters to the top 600 chemical companies that reported releasing such chemicals, asking for their support. More than half of the companies expressed an interest in voluntarily joining the program.

Reducing the production and emission of toxic waste, recycling toxic materials rather than disposing of them, and conducting research to find nontoxic replacements for toxic materials will help save Earth from further damage. [*See also* CLEAN WATER ACT; DIOXIN; HAZARDOUS WASTE; HEAVY METALS POISONING; MEDICAL WASTE; PCBS; RECYCLING, REDUCING, REUSING; SUPERFUND; and TOXIC WASTE, INTERNATIONAL TRADE IN.]

# Toxic Waste, International Trade in

▶ International trade in which groups in some countries get rid of TOXIC WASTE by paying other countries to take the waste. The countries that accept such waste are usually poor and welcome the income. In one highly publicized case, a shipload of toxic waste from Europe arrived in Somalia during the 1992 famine. Because of such situations, some of the worst toxic-waste dumps are in the poorest countries.

Attempts are being made to stop the trade in toxic waste. In the United States, federal laws now regulate many aspects of the transport and dumping of all kinds of waste. International attempts to control toxic substances are also being made. The United Nations has established an International Registry of Potentially Toxic Substances, which provides information about the environmental hazards of many substances. [*See also* HAZARDOUS WASTE and HAZARDOUS WASTE, STORAGE AND TRANSPORTATION OF.]

# Tragedy of the Commons

▶ The idea that when property is owned by a large group of people, some of those people acting in their own interests will overexploit some resources. The "tragedy of the commons" belief first appeared in an essay written in 1883 by William Forster Lloyd. In the essay, Lloyd described how PUBLIC LANDS such as a common pasture land in an English village could be destroyed by overuse. For instance, it seems to be in the self-interest of farmers to add just a few more cows to their herds. Doing this will increase the short-term benefit for each farmer. However, the farmers will suffer in the long run since the land, or commons, will be OVERGRAZED. Lloyd argued that free use of a common resource would lead to the destruction of that resource.

The tragedy of the commons idea is today used to predict the effects of continuing world POPULATION GROWTH. This idea was presented in an article written by Garrett HARDIN that appeared in the journal *Science* in 1968. Hardin feels that belief in "If I don't use it, someone else will" is part of human nature. Because of this belief, Hardin predicted the results of unchecked human population growth. He suggested that such unchecked growth would devastate the human SPECIES because resources such as food, water, and energy would be used up quickly. In addition, living space would become hard to find. People can't increase available supplies of water, energy, and space in a short period of time. Thus, a limit to population growth is needed to avoid resources depletion.

The commons idea, as it relates to the use of NATURAL RESOURCES, has been criticized. Many people object to the idea that public, rather than private, ownership of a resource directly leads to the destruction of the resource. Some critics point to how the timber industry has depleted much of the privately owned OLD-GROWTH FORESTS of the United States. To support their belief, they insist that public FORESTS have survived better than private lands because they are managed for MULTIPLE USE, rather than for one purpose. Thus, if managed properly, there is little chance for overuse of a particular resource.

Some people believe that rare and precious resources can be maintained only through collective ownership of such resources. These people emphasize that it is wrong to assume that the depletion of a resource is related only to population size. Other factors, such as differences in the way a resource is used or distributed, must also be considered. [*See also* ALTERNATIVE ENERGY SOURCES; CONSERVATION; ENVIRONMENTAL ETHICS; FRONTIER ETHIC; OVERGRAZING; RENEWABLE RESOURCES; and SUSTAINABLE DEVELOPMENT.]

# Tree Farming

▶ The creation, growth, and maintenance of artificial FORESTS for the purpose of obtaining wood. Timber companies in the United States often have obtained the trees they need from PUBLIC LANDS such as NATIONAL FORESTS. The removal of trees from such lands is carefully monitored by the U.S. FOREST SERVICE. Because these lands are owned and managed by the federal government, timber companies are restricted in the number and types of trees that can be removed from an area. Restrictions also determine how old or large trees must be before they can be removed. In addition, concern about ENDANGERED SPECIES have led to the elimination or severe restriction of tree removal from some forest areas where heavy logging was once permitted. For example, to preserve the HABITAT of the endangered NORTHERN SPOTTED OWL, timber companies that once logged forests of the Pacific Northwest were forbidden from continuing this practice. To avoid such restrictions and to ensure that timber will be available in the future, many timber companies have turned their attention to the development of tree farms.

Tree farms are privately owned areas used to grow trees that will be harvested mostly for timber or paper production. Tree farmers use many of the same agricultural methods as farmers who grow crops that are to be used as food. For example, when trees are harvested, seedlings are immediately planted to replace those trees that were removed. In addition, fertilizers and PESTICIDES are used to help trees grow more rapidly and in greater numbers than trees on public lands, where their growth is controlled only by nature. The use of tree farms provides timber companies with a continuous supply of wood for harvesting. Because the farms are privately owned, timber companies can also carry out their business without strong regulation from the U.S. Forest Service and the public.

Tree farms provide timber companies with many advantages. However, some people are concerned about how the operation of such farms affects the ENVIRONMENT. One concern is that the use of fertilizers and pesticides on tree farms poses the same threats to the environment as do the use of these substances on conventional farms. In addition, many tree farms make use of MONOCULTURE farming—the growth of only one plant SPECIES. Because all trees grown on a monoculture farm are alike, they have the same characteristics. As a result, such crops may be more vulnerable to NATURAL DISASTERS, such as droughts, diseases caused by microorganisms, FUNGI, and INSECTS. [*See also* AGROFORESTRY; DEFORESTATION; and HABITAT LOSS.]

◆ Private timber companies often create tree farms such as this one to obtain the wood they need for timber and paper production.

# Trophic Level

▶ A layer in the structure of feeding relationships in a FOOD CHAIN. Simply stated, a trophic level is a link in a food chain. Food chains are simple models scientists use to show how nutrients and energy move through an ECOSYSTEM. As organisms eat one another, energy and nutrients are transferred from one trophic level to the next.

## TROPHIC LEVELS IN FOOD CHAINS

Food chains usually consist of three to five trophic levels. An organism's NICHE determines which trophic level it feeds on. For instance, PRODUCERS, such as PLANTS and ALGAE, always occupy the first trophic level in a food chain. That's because producers—organisms that manufacture food energy from sunlight or chemicals in the ENVIRONMENT—make energy available to all other organisms in the ecosystem. Producers are also known as AUTOTROPHS, which means "self-feeders." Thus, these organisms "feed" on the first trophic level. Without the critical work of producers, life would not be possible.

Occupying the second trophic level are HERBIVORES, organisms that feed on producers. Familiar herbivores include grass-eating cattle, algae-eating FISH, and leaf-eating GORILLAS. As these organisms eat, they obtain some of the energy and nutrients stored in the producers. Organisms on the second trophic level are also known as first-order CONSUMERS because they are the first consumers in the food chain sequence.

Second-order consumers, or CARNIVORES, occupy the third trophic level in an ecosystem. These are the meat-eaters, which feed on first-order consumers. Examples of second-order consumers include the heron, a carnivorous BIRD that feeds on fish, frogs, and other small animals, and foxes, which feed on rabbits, mice, and many other small herbivores.

Occupying the fourth trophic level are third-order consumers, carnivores that feed on second-order consumers. Third-order con-

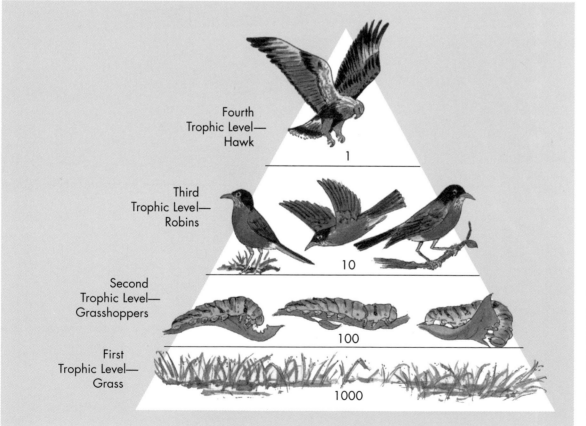

Fourth Trophic Level— Hawk

1

Third Trophic Level— Robins

10

Second Trophic Level— Grasshoppers

100

First Trophic Level— Grass

1000

◆ This energy pyramid shows how energy is lost from one trophic level to the next. Grass plants on the first trophic level store 1,000 times more energy than the hawk at the fourth trophic level.

sumers, such as killer WHALES, many types of sharks, and alligators, are usually the top PREDATORS in an ecosystem.

## TROPHIC LEVELS AND THE TRANSFER OF ENERGY

Scientists assign organisms to different trophic levels in order to study how energy is transferred through an ecosystem. To visualize the passage of energy, scientists have developed a model called an ENERGY PYRAMID. An energy pyramid shows how energy is lost from one trophic level to the next. In general, the total energy transfer from one trophic level to the next is only about 10%. What happens to the other 90% of the energy? Organisms use much of their energy (about 90%) to carry out the functions of living—moving around, growing, finding food, and so on. This energy is converted into heat and is lost from the food chain. Thus, when an organism is eaten, that energy is not available anymore—it has already been used.

The energy pyramid is a simple model, but the feeding relationships within ecosystems can be much more complex. SPECIES may feed at several trophic levels. For instance, OMNIVORES, such as humans, eat producers and consumers. DECOMPOSERS, such as FUNGI, protists, and BACTERIA, can feed on any trophic level because they break down and decay the bodies of dead organisms. [*See also* BIOGEOCHEMICAL CYCLE; BIOLOGICAL COMMUNITY; BIOSPHERE; DECOMPOSITION; ECOLOGY; and FOOD WEB.]

---

## Tropical Rain Forest

*See* RAIN FOREST

---

## Tropics

▶ The areas of Earth close to the **equator**, which are characterized by warm WEATHER. The average annual temperatures in the tropics range from 77° F to 90° F (25° C to 32° C).

The tropics extend on both sides of the equator, from $23\frac{1}{2}°$ north **latitude** to $23\frac{1}{2}°$ south latitude. The northern limit of the tropics is a hypothetical line known as the *Tropic of Cancer*. The southern limit is a hypothetical line called the *Tropic of Capricorn*.

◆ The tropics are characterized by a year-round warm climate.

## THREE MAJOR CLIMATE AREAS

The tropics include three major CLIMATE areas—rainy tropics, tropical SAVANNAS, and DESERTS. The rainy tropics, also known as tropical RAIN FORESTS, are located mostly in the lowlands close to the equator and along the coasts. They are characterized by heavy PRECIPITATION, averaging more than 80 inches (203 centimeters) per year.

The tropical savannas are a little farther away from the equator than the rainy tropics. Precipitation is not as heavy, and there is a long dry season.

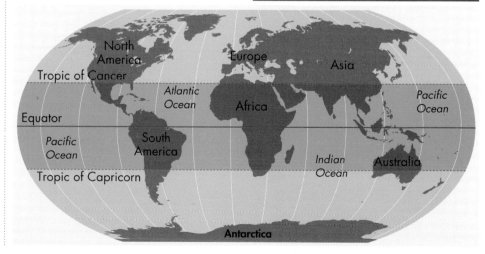

Tropical deserts are the tropical areas located farthest away from the equator. The average annual precipitation is very low—less than 10 inches (25 centimeters).

## THE POSITION OF THE SUN

Unlike areas of higher latitude, such as the United States, the tropics have only small seasonal temperature changes. Why is this? In the tropics, the sun is almost directly overhead, and its rays strike Earth directly. The incoming rays are concentrated on a small area of Earth's surface. Tropical areas receive more SOLAR ENERGY than do other parts of the world, which is why it is always warm in the tropics.

The sun's position also has a direct effect on the amount of rainfall in the rainy tropics. When solar energy hits the ground, the air above it is warmed. Warm air can hold a large amount of water vapor. As the warm air rises, it slowly cools and loses some of its ability to hold water. When the air is cooled to a certain point, water vapor is converted to liquid water, which then falls as rain.

## RAIN FOREST BIODIVERSITY

The rainy tropics contain the world's greatest BIODIVERSITY. Scientists have identified nearly 1.5 million different SPECIES of PLANTS, animals, and other organisms on Earth. Over half of these are species inhabiting the tropical rain forests of South America, Africa, and Southeast Asia.

The warm, wet climate is ideal for many plants. There are more species of plants growing in tropical rain forests than in any other BIOME on Earth. One hundred acres (40 hectares) of tropical rain forest may contain 100 different species of plants, compared to 10 species in the same area of temperate forest.

Scientists are concerned that disruptions to tropical rain forests may cause a loss of species. DEFORESTATION poses the greatest threat. It is estimated that between 16,000 and 80,000 square miles (between 41,000 and 210,000 square kilometers) of tropical rain forest is lost annually from deforestation. [See also CLIMATE; ENDANGERED SPECIES; EVAPOTRANSPIRATION; EXTINCTION; HABITAT LOSS; SAVANNA; and SPECIES DIVERSITY.]

# Troposphere

The layer of the ATMOSPHERE nearest Earth's surface. The troposphere extends from Earth's surface to an altitude of 10 to 12 miles (16 to 19 kilometers) at the equator. At the poles, the troposphere extends from Earth's surface to an altitude of 5 to 6 miles (8 to 9.7 kilometers). Almost all human activity, the WATER CYCLE, and nearly all Earth's WEATHER take place in the troposphere.

About four-fifths of the mass of the atmosphere is contained in the troposphere. As much as 99% of the water vapor and CARBON DIOXIDE in the atmosphere is also found in this layer. Both these gases are important in maintaining the temperature balance of Earth.

In the troposphere, temperature decreases as altitude increases. The decrease in temperature averages about 18° F for each mile (−7.8° C for each 1.6 kilometer) up. Temperature begins to level off at about −67° F (−55° C) near the upper boundary of the troposphere. This area of the troposphere is called the *tropopause*. These great temperature differences within the troposphere cause the air in this layer to be heated unevenly. Denser, cooler air at higher altitudes tends to sink toward Earth's surface. As the cool air sinks, it is heated by the sun-warmed ground and rises again. This movement of air forms vertical currents. Earth's rotation produces horizontal currents in the upper troposphere. These currents, called *jet streams*, move air from west to east.

Because the air in the troposphere is in constant motion, weather processes influence the way AIR POLLUTION is carried from place to place. Pollutants tend to be confined to one area until local winds and PRECIPITATION carry or wash them away. At times, such as during thunderstorms, pollutants may be carried rapidly upward by vertical currents in the troposphere. At altitudes greater than half a mile (0.8 kilometers), they may be picked up by the jet stream. The jet stream can reduce the effects of the pollutants and spread them over an extended area. Thus, emissions from cars and other machines that burn FOSSIL FUEL may be diluted or concentrated in the troposphere, affecting the formation of such conditions as ACID RAIN and photochemical SMOG. [See also CARBON

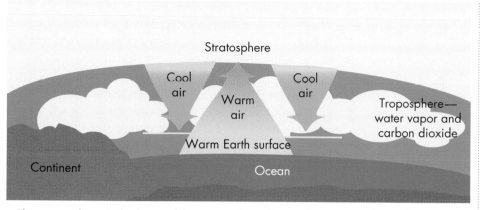

The troposphere is the layer of the atmosphere that is closest to Earth's surface.

MONOXIDE; GLOBAL WARMING; HYDRO-SPHERE; MESOSPHERE; NITROGEN DIOXIDE; NITROGEN OXIDE; OZONE; RADIATION; SOLAR ENERGY; STRATOSPHERE; and SULFUR DIOXIDE.]

# Tuna

▮Large, fast-swimming SPECIES of saltwater FISH. Tuna are a popular game fish and food fish. They are sold canned, fresh, and frozen as food in many countries. Japan, the Philippines, and the United States are the main consumers of tuna meat. According to the Food and Agriculture Organization of the United Nations, Japan is the single largest consumer of tuna meat.

The 14 species of tuna are widely distributed in the temperate and tropical OCEAN waters of the world. They range as far north as Newfoundland and Norway in the Atlantic Ocean and northern Japan in the Pacific Ocean. Tuna are known to migrate extremely long distances. In one study, scientists discovered that one fish that was tagged and released off Japan had migrated 6,700 miles (10,720 kilometers) to Mexico.

Tuna are caught in several ways. A chief method of capturing tuna is called *purse seining*. In this technique, a large net is used to encircle and trap schools of tuna. Purse seining has been severely criticized by environmental activists because DOLPHINS, which often chase large schools of tuna, sometimes get caught in the nets as well.

Due to public pressure, major producers of canned tuna in the United States announced in 1990 that they would no longer use tuna caught in nets that also trap dolphins. Today, the United States and other countries require purse seining nets to have escape chutes so that dolphins can swim freely through them. Thus, canned tuna producers now show the dolphin-safe logo on their products. [*See also* FISHING, COMMERCIAL; and MARINE MAMMAL PROTECTION ACT.]

Public outcry over the unintentional trapping of dolphins in tuna nets prompted the U.S. tuna industry to use nets that have escape chutes for dolphins.

◆ The fourteen species of tuna range from Newfoundland and Norway in the Atlantic Ocean and northern Japan in the Pacific Ocean. They migrate long distances.

# Tundra

▌▶A nearly treeless BIOME having a CLIMATE and vegetation characteristic of Arctic regions and high mountain areas above the **treeline**. The tundra covers almost one-third of Earth's surface. It is the predominant biome of the northern half of Alaska, parts of Canada and Greenland, and the northern regions of Scotland, Ireland, Scandinavia, China, and Siberia.

There are basically two climates in the tundra: a mild, springlike summer and a long, cold, dark winter. Average low temperatures in the tundra are about −25° F (−32° C) in winter. In many areas, the deep layers of tundra SOIL remain frozen throughout the year. This permanently frozen soil, called *permafrost,* prevents rain and water from melting and prevents them from draining into soil layers. Total PRECIPITATION in the tundra averages between 4 and 12 inches (10 and 30 centimeters) per year. This is a little more than most of the world's DESERTS.

Although the sun barely sets during the tundra summer, the season is short and cool. Temperatures reach a high of about 40° F (4° C). These temperatures are just high enough to thaw the layer of soil above the permafrost. However, the winds may be so strong (even in the summer) that the temperature 1 foot (0.3 meters) above the ground may be much cooler than at ground level.

## A FRAGILE ECOSYSTEM

During the summer thaw, water remains trapped near the ground surface. This trapped water forms stretches of **bogs**, lakes, and rivers. Tundra soil tends to be acidic and low in nitrogen, and has little bacterial activity. As a result of these factors, DECOMPOSITION is slow. As organic matter accumulates on the ground, the soil becomes waterlogged, further slowing decomposition.

Only low, ground-hugging PLANTS can survive in the tundra. The harsh winds, low temperatures, short growing season, and permafrost layer prevent the growth of large plants. Mosses, LICHENS, grasses, shrubs, and a few trees that are dwarf in size dominate the landscape. These plants form the base of tundra FOOD CHAINS and FOOD WEBS.

Although few SPECIES can survive in the tundra, those that do exist in great numbers. Reindeer moss provides food for vast herds of caribou (in North America) and reindeer (in Europe and Asia) that wander through the tundra. Snowshoe hares, lemmings, and ptarmigans graze on the low-lying plants. Wolves and arctic foxes, as well as polar and GRIZZLY BEARS, are the major PREDATORS of the tundra. Among the many INSECT species that inhabit the tundra, black flies, deerflies, and mosquitoes that breed in tundra bogs and marshes provide food for migrating BIRDS that visit the tundra each summer.

## SOME THREATS TO THE TUNDRA

Because of the slow decomposition rate, shallow soil layer, and short growing season, the tundra takes a long time to recover when it is harmed or damaged. Oil exploration, MINING of gold and COPPER, road building, and other human activities change the tundra surface. When the thin layer of soil and vegetation that covers and insulates the permafrost is removed or destroyed, the depth of the summer thaw is extended. Melting snow and ice mixes with the silt below, forming mud that is easily eroded and carried away by meltwater streams. Precious TOPSOIL is lost.

Changes in the amount of atmospheric and terrestrial CARBON in tundra regions could signal the beginning of global climatic changes. GLOBAL WARMING may result in the thawing of permafrost. In addition to EROSION, the thawing of permafrost would allow large amounts of METHANE, now locked in ice, to be released into the ATMOSPHERE.

The summer growing period in the tundra is less than four months long. This means that plants have a limited time to seed and reproduce. Any CLIMATE CHANGES could damage a species' reproductive ability for several generations, affecting all levels of the tundra food web. [*See also* ALASKA NATIONAL WILDLIFE REFUGE (ANWR); ALASKA PIPELINE; BIOMASS; CLIMATE CHANGE; EXXON VALDEZ; and MIGRATION.]

# U

## Ultraviolet Radiation

▌One type of RADIATION given off by high-temperature objects, such as the sun. Radiation is a type of energy that travels through space in the form of waves. Ultraviolet

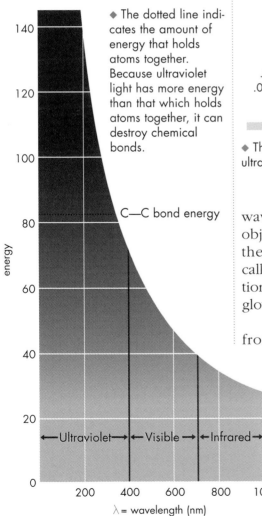

◆ The dotted line indicates the amount of energy that holds atoms together. Because ultraviolet light has more energy than that which holds atoms together, it can destroy chemical bonds.

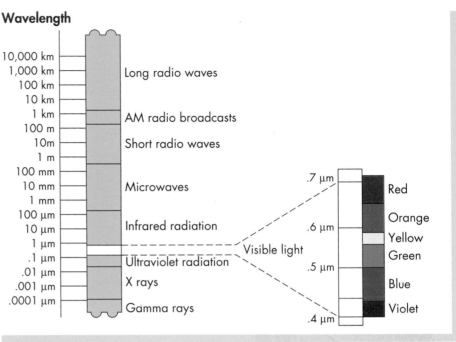

◆ The diagram above is the electromagnetic spectrum showing the wavelength of ultraviolet light.

waves cannot penetrate deep into objects and therefore affect only their outer surfaces. Sometimes called *black light*, ultraviolet radiation can cause certain surfaces to glow and emit light.

Ultraviolet radiation coming from the sun is important for human health. Ultraviolet light causes skin cells to produce vitamin D, an important substance in bones and teeth. Ultraviolet light also causes tanning of the skin by activating pigments, chemicals that produce coloring in

skin cells. Tanning is a natural body defense that helps protect skin from injury caused by sunlight. However, exposure to ultraviolet radiation is known to cause several human health problems. One result of too much exposure to ultraviolet light is a painful reddening of the skin known as sunburn. Long-term exposure to the sun is even more serious. Ultraviolet radiation contains a great deal of energy. When it strikes biological compounds in the skin, it can break chemical bonds that hold the atoms of these compounds together. People who are constantly exposed to the sun, such as farmers, construction workers, and lifeguards, are much more at risk of developing skin disorders, such as thickening of the skin, rapid aging, and CANCER.

Too much ultraviolet radiation can affect other living things as well. For instance, overexposure to ultraviolet light can kill microscopic ALGAE, called PHYTOPLANKTON, that live in the OCEANS. The death of phytoplankton disrupts PHOTOSYNTHESIS and reduces the amount of OXYGEN given off by the algae. Death of algae also disrupts ocean FOOD CHAINS and FOOD WEBS.

## ULTRAVIOLET RADIATION AND THE OZONE LAYER

Life on Earth is protected from overexposure to ultraviolet radiation by a natural sunscreen in the STRATOSPHERE. Known as the OZONE LAYER, this portion of the ATMOSPHERE absorbs some of the ultraviolet radiation coming from the sun. In doing so, the ozone layer shields Earth's inhabitants from the sun's dangerous rays.

Since the 1970s, scientists have been concerned about a group of artificial chemicals called chlorofluorocarbons (CFCs). CFCs damage the ozone layer. If the ozone layer becomes damaged, more ultraviolet light can reach Earth, placing living things at risk of overexposure. Since the 1970s, the United States and other industrialized nations have passed laws to limit the use and production of CFCs and other substances that may cause damage to the ozone layer. [*See also* AIR POLLUTION; CARCINOGEN; CLEAN AIR ACT; ELECTROMAGNETIC SPECTRUM; MONTREAL PROTOCOL; OZONE; OZONE HOLE; and RADIATION EXPOSURE.]

# United Nations Conference on the Human Environment

▌Conference held in 1972 to discuss ways in which member nations of the United Nations (U.N.) could work together to protect the ENVIRONMENT. In 1968, the General Assembly of the United Nations unanimously agreed to hold a conference because of the "continuing and accelerating impairment of the quality of the environment" due to POLLUTION, EROSION, wastes, and the secondary effects of biocides, or substances destructive to organisms.

The conference took place in Stockholm, Sweden, June 5 to June 16, 1972. There were three parts to the work of the conferees: to produce a "Declaration on the Human Environment" stated in 26 principles; to produce an action plan; and to establish new U.N. agencies to oversee the actions. The UNITED NATIONS ENVIRONMENTAL PROGRAMME (UNEP) was established with headquarters in Nairobi, Kenya, headed by Maurice Strong of Canada. As a result, such efforts as Earthwatch were established to monitor global air, water, and land conditions.

Since 1972, the United Nations has held conferences on specific aspects of environmental problems: Population and Food in 1974; many Law of the Sea conferences; Habitat in 1976; Water in 1977; Population in 1984; Earth Summit in 1992; Population and Development in 1994; and Climate Change in 1995.

# United Nations Earth Summit

▌A meeting held by the United Nations in 1992 in Rio de Janeiro, Brazil, to address global environmental policies. Officially known as the United Nations Conference on Environment and Development, it was the largest environmental meeting ever held, with nearly 180 nations participating. The purpose of the summit was to address such issues as BIODIVERSITY, GLOBAL WARMING, SUSTAINABLE DEVELOPMENT, and relationships between developed and developing nations that relate to environmental matters.

Major outcomes of the Earth Summit included:

• The Rio Declaration on Environment and Development—a statement of principle governing worldwide policy on the ENVIRONMENT and development

• Agenda 21—an action plan for environmental protection and sustainable development

• Biodiversity Convention an agreement signed by all participants except the United States that calls

◆ U.S. Vice President Albert Gore, Jr., speaks at the U.N. Earth Summit held in Rio de Janeiro, Brazil, in 1992.

funds will be made available to help those nations.

## EARTHWATCH

Through project Earthwatch and its foundation, the Global Environmental Monitoring System (GEMS), the UNEP acquires data about world environmental problems to provide early warning of any impending danger to the international community and to ensure that those problems receive quick and appropriate action. GEMS, which went into action in 1975, has run many major projects, including the monitoring of CLIMATE CHANGES, health problems, long-range transportation of pollutants, and ocean activity. In 1984, GEMS assessed damage to the OZONE LAYER, and in 1985 it set up a world conference on chlorofluorocarbons (CFCS).

## ORIGIN OF THE UNEP

In December 1968, the U.N. General Assembly passed a resolution

on nations to list SPECIES to be preserved and to develop strategies to conserve and use biological resources (the United States refused to sign because of language deemed unacceptable)

• Global Warming Convention—a commitment by nations to reduce greenhouse emissions

• Statement on Forest Principles—a statement relating to the protection of the world's FORESTS

preserve Earth's NATURAL RESOURCES. The United Nations Environmental Programme (UNEP) monitors world pollution problems and helps world communities find solutions. In addition, the UNEP reviews the effects any international environmental projects might have on developing nations, such as additional financial costs, to ensure that resources or

# United Nations Environmental Programme (UNEP)

❙▶A program launched by the United Nations (U.N.) in 1972 to encourage international cooperation in the effort to stop POLLUTION and

◆ The majority of the United Nations delegates in the General Assembly approved the establishment of the United Nations Environmental Programme to monitor the world's environmental problems and encourage international cooperation to solve the problems.

pledging to work toward solving environmental problems. The UNITED NATIONS CONFERENCE ON THE HUMAN ENVIRONMENT was held in 1972 to develop ways in which member nations could cooperate to protect the ENVIRONMENT. After the conference, plans for the establishment of the United Nations Environmental Programme were presented to the assembly in a resolution sponsored by 19 nations: Argentina, Brazil, Canada, Cyprus, Ghana, Greece, Guatemala, Iran, Jamaica, Kenya, Malta, Mexico, New Zealand, Nigeria, Swaziland, Sweden, Tanzania, Tunisia, and the United States. The assembly adopted the resolution 116 to 0, with ten delegates not voting, and agreed to set up headquarters in a developing nation. On January 1, 1973, the UNEP officially went into action from an office in Nairobi, Kenya.

## HELPING MEMBER NATIONS HELP THE ENVIRONMENT

Education and training in areas such as nature CONSERVATION, control of OIL POLLUTION, and the use of chemical dispersants are available through the UNEP to both governmental and nongovernmental organizations. The Environmental Liaison Centre in Nairobi developed a network of 6,000 nongovernmental groups that deal with environmental issues. Through the center, interested parties can receive information on a variety of subjects, including worldwide atmospheric and oceanographic research, environmental awareness and conservation, ECOSYSTEMS, water sources, and DESERTIFICATION. Films, videotapes, and more than 10,000 color photos are lent to organizations that promote ecological awareness. Such films as *The State of the Planet* and *Water: A Vital Resource* are available in a variety of languages. In 1984, the UNEP shot *Seeds of Despair*, a film about desertification in Ethiopia. The film focused global attention on Ethiopia's FAMINE and brought about worldwide fundraising efforts to help the starving people there. [*See also* ENVIRONMENTAL IMPACT STATEMENT and UNITED NATIONS EARTH SUMMIT.]

# Uranium

▶ A radioactive element used as a FUEL to produce NUCLEAR POWER. Uranium is a naturally occurring substance in Earth's crust. Uranium, PLUTONIUM, and other radioactive substances are used as sources of nuclear fuel because they contain many times more energy than FOSSIL FUELS such as COAL or oil.

Although uranium was discovered in 1789, it did not become the subject of intense study until 1938, when scientists discovered that it could be used in NUCLEAR FISSION reactions to produce energy. This and other scientific discoveries led to the use of uranium in the first nuclear-powered electrical generator in the United States in 1957.

Today, uranium is the most common nuclear fuel because it can produce tremendous amounts of energy. One pound (0.45 kilogram) of uranium can produce as

◆ Uranium can produce enormous amounts of energy.

much energy as 3 million pounds (1.36 million kilograms) of coal. Uranium is mined in many parts of the world, including North America, Europe, Africa, and Australia.

However, uranium does have disadvantages. Perhaps the most serious is that uranium and other nuclear fuels produce nuclear waste. Another disadvantage is that easily obtainable supplies of uranium on Earth are decreasing and the costs of locating and refining it are extremely high. [*See also* ALTERNATIVE ENERGY SOURCES; BREEDER REACTOR; INTERNATIONAL ATOMIC ENERGY AGENCY (IAEA); MINING; NUCLEAR WINTER; RADIATION; RADIATION EXPOSURE; RADIOACTIVE FALLOUT; and RADIOACTIVITY.]

# V

## Vertebrate

▶ An animal with a backbone and an internal skeleton made of bone or cartilage. The backbone is a structure made up of bony or cartilaginous segments called *vertebrae*. It is from these structures that the group receives its name. Vertebrates include all animals classified as FISH, AMPHIBIANS, REPTILES, BIRDS, and MAMMALS.

### CHARACTERISTICS OF VERTEBRATES

Besides a backbone, vertebrates also share other ADAPTATIONS, including a skeletal system made up of separate bones, a muscular system that helps move the bones, and a complex nervous system that controls the internal processes and external actions of the organism.

### Skeletal System

Skeletons are not unique to vertebrates. Other types of animals also have internal and external skeletons. For example, all arthropods, an INVERTEBRATE group that includes shrimp, lobsters, crabs, spiders, and INSECTS, have an external skeleton made of chitin. Echinoderms, an invertebrate group that includes sea stars and sea urchins, have internal skeletons made up of calcium carbonate. None of these animals, however, have backbones.

The skeleton of most vertebrates is made up of a skull, a backbone or vertebral column, a rib cage, and two pairs of limbs. This skeleton serves three important functions:

**1.** it supports the body and provides areas for attachment of muscles;

**2.** it protects internal organs, such as the heart, lungs, and brain; and

**3.** it helps store MINERALS, such as calcium and phosphorus, that are needed by the organism.

Another important skeletal adaptation in vertebrates is the EVOLUTION of jaws. All vertebrates, except for sea lampreys and their close relatives, have jaws made up of upper and lower sections. Jaws enable an animal to grasp and hold on to prey. Teeth-bearing jaws also allow animals to chew and feed on a variety of different food items.

### Muscular System

Vertebrates have muscular systems that allow them to carry out a variety of movements. Skeletal muscles are made up of thick bundles of cells that attach at different points on the skeleton. Movement of the skeleton occurs when a muscle contracts. The vertebrate skeleton also has many joints, areas where two bones meet. This allows vertebrates to move their heads, trunk, and limbs into many different positions.

### Central Nervous System

Vertebrates have a complex central nervous system. A central nervous system is made up of a brain and a spinal cord. Nerves branch off the spinal cord and into the skin, bones, muscles, and internal organs.

The vertebrate nervous system is highly efficient. For example, the brain is divided into different regions that have different functions. This enables vertebrates to have greater sensitivity to images, sounds, odors, and vibrations than invertebrates. It also allows vertebrates to engage in more complex behaviors than other animals.

### TYPES OF VERTEBRATES

Approximately 40,000 SPECIES of vertebrates have been identified. Scientists classify or place these species into groups according to similarities in structure. Biologists recognize five different classes of vertebrates: fish, amphibians, reptiles, birds, and mammals.

### Fish

Fish are found living in nearly every aquatic HABITAT on Earth. There are three different types of fish: the jawless fish, which includes the

**cartilaginous fish** fish whose skeleton is entirely or largely composed of cartilage.

◆ The blue poison dart frog is an amphibian found in South American forests.

◆ Fish are the most abundant vertebrates.

lamprey and hagfish; the **cartilaginous fish**, which includes sharks and rays; and the bony fish, which includes SALMON, trout, TUNA, and all other fish. With approximately 20,000 living species, fish are by far the most abundant vertebrates. In fact, there are more different types of fish than all other vertebrates combined.

Fish come in a variety of shapes and sizes, from the tiny dwarf goby which measures less than 0.4 inches (1 centimeter) in length, to the huge whale shark, which can reach lengths of up to 50 feet (15 meters). Although they are an extremely diverse group, fish share many similarities, including gills for obtaining DISSOLVED OXYGEN from the water and fins to help steer and swim.

## Amphibians

Amphibians include frogs, toads, salamanders, and caecilians. In some ways, amphibians are an intermediate group between fish and reptiles. All amphibians change in form as they develop. For example, they begin life as aquatic larvae, such as tadpoles, and change into air-breathing adults that can spend at least part of their time on land. This striking change, or meta-

morphosis, gives the class Amphibia, which means "double life," its name.

Amphibians are a diverse group that contains about 2,400 species. They are distributed worldwide. However, amphibians most often live in regions that have warm temperatures all year. Amphibian eggs lack protective coverings and must be laid in water or moist areas. Thus, a supply of water is also needed for breeding.

## Reptiles

The class Reptilia includes snakes, lizards, turtles, and alligators. This diverse group contains nearly 6,000 living species. Like amphibians, most reptiles live in regions having warm temperatures. A warm climate is important because amphibians and reptiles are ectotherms. Ectotherms are animals that rely on sunlight or a warm ENVIRONMENT if they need to raise their body temperature. Because ectotherms do not have to generate body heat, as birds and mammals do, they are much more efficient at food processing so that it goes directly into body growth; consequently, they can survive on far less food.

Most reptiles lay eggs that have tough, leathery shells. These shells protect the developing embryos from injury and dehydration in a land environment. Other reptiles give birth to live young.

◆ Reptiles like this thorny devil lizard live in warm climates.

## Birds

With nearly 9,000 living species, birds are the second largest group of vertebrates. Birds have one adaptation no other animals have—feathers. Feathers help birds maintain their body temperature. Feathers also give most birds the ability to fly.

Other animals, such as flying fish and squirrels, can glide for short distances. However, birds and bats, which are mammals, are the only vertebrates capable of sustained flight. Birds possess a number of adaptations for flight. These adaptations include wings; a small body size; large breast muscles; lightweight, hollow bones; and lightweight beaks.

◆ The flamingo's feathers help it to maintain its body temperature.

◆ The olive baboon is a mammal found in Africa. Like all mammals, it nourishes its young with milk secreted by mammary glands.

## Mammals

Mammals, such as dogs, DOLPHINS/ PORPOISES, ELEPHANTS, monkeys, and humans, are vertebrates that have fur or hair. Unlike other vertebrates, few mammals lay eggs. Instead, mammalian embryos develop for some period of time inside the mother and are born alive.

Mammals get their name from the mammary glands of female mammals. Mammary glands secrete milk. The milk is used to nourish offspring until they are old enough to find food and protect themselves.

Like birds, mammals are animals that maintain constant body temperatures regardless of the temperature of their environment. This adaptation has enabled birds and mammals to inhabit virtually every type of environment on Earth.

# Virus

A microscopic particle consisting of deoxyribonucleic acid (DNA) or **RNA** surrounded by a **protein** coat. Viruses cause a variety of diseases, including mumps, measles, influenza (the "flu"), polio, and AIDS (acquired immunodeficiency syndrome). Although viruses cause disease as some living organisms do, most biologists do not consider viruses to be alive. That is because viruses do not carry out all life processes. For instance, viruses do not use food for energy, they do not grow, and they do not move around on their own. Viruses can only reproduce, and they cannot even do this by themselves. Viruses can only reproduce inside living cells.

Viruses infect many types of organisms. For instance, more than 400 different viruses are known to infect PLANTS. In fact, the first virus ever to be identified was the tobacco mosaic virus, which causes yellow spotting on the leaves of the tobacco plant. Most types of VERTEBRATES can also contract viral diseases, as can many species of arthropods, such as crabs, shrimp, and INSECTS. However, viruses are not common in other animals.

Because viruses are not alive, they cannot be killed by medicines, such as antibiotics. This is why medicine cannot help you get rid of the flu the way it might help you get rid of pneumonia or skin infections caused by BACTERIA. Instead, especially for extremely dangerous viruses, a person must receive a vaccine, which helps the body's own immune system fight off the virus before it can cause the disease. [*See also* FLOWERING PLANT; HEALTH AND DISEASE; and PATHOGEN.]

---

### THE LANGUAGE OF THE ENVIRONMENT

**protein** a complex chemical essential to all life; protein is important for muscle contraction, oxygen transport, and many other cell activities.

**RNA** ribonucleic acid; RNA is a chemical cousin of DNA and is important for many cell activities.

---

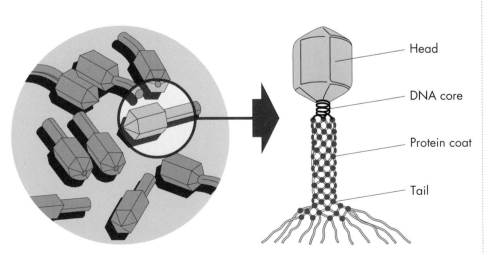

Head
DNA core
Protein coat
Tail

◆ Viruses are smaller than most living cells and come in a variety of shapes and sizes. Each virus is specifically adapted for the type of cell it infects.

# Volcanism

The forceful discharge of magma, rock fragments, gases, and water through an opening in the surface of Earth. Liquid rock is called *magma* and has a temperature range of 1500° to 2700° F (816 to 1482° C). Magma on the surface is called *lava*. Magma, rocks, and ash can be expelled with a force equal to many atomic bombs.

There are several forms of volcanism. In one form, rock debris is ejected with great force from a central vent of a cone-shaped landform. In another, molten rock pours out of cracks that occur along a long fracture or fractures. In a third kind, groundwater heated by buried magma spouts from a geyser or as steam from a vent called a *fumarole*.

The Earth's surface, which includes the bottom of OCEANS as well as the continents, is made of continental plates whose movements and behavior are studied in a discipline called PLATE TECTONICS. The plates are huge slabs of rock that tend to move due to move-

◆ The eruption of Mount Saint Helens in 1980 was one of the largest in North American history.

ments below them inside of Earth. During Earth's history, the plates have moved around a great deal. When populations of organisms that live on different plates become isolated due to plate movement, EVOLUTION of new SPECIES is made more likely because such species are also genetically isolated.

There are about 800 known active volcanoes. Most are in the "circle of fire" that rings the Pacific. Others are related to the worldwide rift system, which includes mid-ocean ridges and continental rifts such as those in East Africa. Eruptions of oceanic volcanoes may result in the creation of new islands. The Hawaiian Islands originated in repeated eruptions of volcanic material accumulating at the top of a submarine platform or ridge.

Volcanoes cause great damage and destruction, ranging from environmental changes to human deaths. In 1980, Mount Saint Helens erupted, spewing ash. All trees were destroyed for 8 miles (13 kilometers). Ash was thrown as high as 4 miles (6.5 kilometers) up into the ATMOSPHERE and traveled around Earth, interfering with sunlight and making the land darker and cooler. Mount Saint Helens is one of five major volcanoes located in a line through Oregon and Washington. The line is where one continental plate is sliding under another.

The 1991 eruption of Mount Pinatubo in the Philippines created a sun-reflecting haze that reduced the amount of sunlight striking Earth. Earth was cooled substantially for about two years.

Volcanic eruptions have killed many people. In 1815, an eruption in Tambora, Indonesia, killed 12,000, and then 80,000 more died from the resulting famine. In 1883, an eruption of Krakatoa in Malaysia caused a sea wave that killed 36,000 people. [*See also* CONTINENTAL DRIFT; GEOTHERMAL ENERGY; and LITHOSPHERE.]

# Waste Management

▮Methods by which people dispose of discarded materials, including GARBAGE and hazardous materials resulting from industial and MINING processes. Since the INDUSTRIAL REVOLUTION, the combined effects of a growing human population, increased industrial activity, and introduction of many new manufactured products have made it difficult to find adequate space and means for the disposal of waste materials.

Prior to and shortly after the Industrial Revolution, people generally disposed of unwanted materials in open dumps or in bodies of water such as lakes, rivers, and OCEANS. All these methods of disposal had unwanted side-effects. These side-effects included foul odors; breeding grounds for disease-carrying organisms such as rats, BACTERIA, and INSECTS; and POLLUTION of Earth's land, water, and air. To help eliminate such problems, many communities and industries began to seek other methods to dispose of their wastes.

## DISPOSAL OF SOLID WASTES

In recent years, the most commonly used method for SOLID WASTE disposal has involved the use of LAND-FILLS. In this process, garbage is buried below Earth's surface. However, the use of landfills is not with-

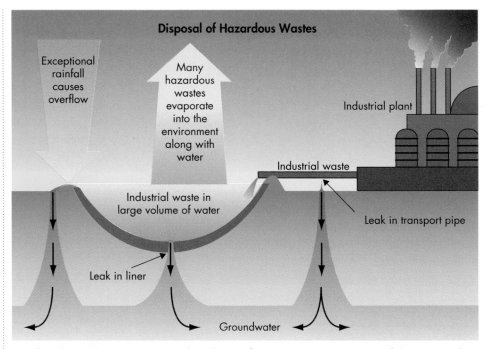

**Disposal of Hazardous Wastes**

Exceptional rainfall causes overflow

Many hazardous wastes evaporate into the environment along with water

Industrial plant

Industrial waste

Industrial waste in large volume of water

Leak in transport pipe

Leak in liner

Groundwater

◆ When hazardous wastes are placed in surface sites, contamination of the surrounding area may occur in several ways, such as leaks, overflows, and evaporation.

out problems. People living near these sites complain about the odors given off by decaying garbage. When it was discovered that rainwater washing through these sites also carried pollutants into nearby bodies of water as well as groundwater supplies, an alternative means of disposal was developed—the sanitary landfill.

In a sanitary landfill, clay and plastic liners are placed in a site before garbage is buried. When the garbage reaches a certain height, it is covered by layers of SOIL. New garbage is then added and again covered with soil until the landfill reaches a predetermined height. Then the site is covered with still

more soil and is often planted with grasses, shrubs, and trees.

Like their predecessors, sanitary landfills are not without problems. The first problem is where to place a landfill. Desired areas should not be located near bodies of water or near groundwater supplies. In addition, a community must usually agree to the construction of the landfill—a task that is not easy to accomplish. Such landfills are expensive to maintain. Special materials to construct liners and constant addition of soil layers are required. Finally, the replanting of the site involves the use of much heavy machinery and requires many workers.

## ALTERNATIVES TO LANDFILLS

In some areas, especially urban centers, people make use of incineration to get rid of their disposed materials. During incineration, refuse is burned in large containers. Incineration reduces large piles of garbage to smaller piles of ash. Unfortunately, incineration is not problem-free. The pile of ash left behind by burned refuse must be disposed of in some manner. Such ashes may contain toxic materials that are harmful to people and other organisms. In addition, the incinerators release smoke and pollutants into the ATMOSPHERE through their smokestacks.

COMPOSTING is used by some people for the disposal of BIO-DEGRADABLE wastes. Such wastes usually include food products, leaves that have fallen from trees, and grass clippings. Composting is helpful to the ENVIRONMENT because through the process of DECOMPOSI-TION, nutrients that are used by PLANTS are returned to the environment. However, composting has two major drawbacks. The first is that compost piles, like open dumps, frequently give off foul odors as decomposition takes place. In addition, composting is limited in use, because only bio-degradable materials can be disposed of in this way.

## HAZARDOUS WASTE DISPOSAL

Many household products and industrial wastes are classified as HAZARDOUS WASTES. Such wastes are difficult to dispose of in a way that will not harm people or the environment. To help lessen the likelihood of harm, the disposal of solid wastes is heavily regulated by individual states and by the federal government. Enforcement of such regulations usually falls to the U.S. ENVIRONMENTAL PROTECTION AGENCY (EPA). Some methods for the disposal of hazardous wastes that are being used include the use of secure landfills, deep-well injection, SOLID WASTE INCINERATION, and resource recovery and waste exchange.

### Secure Landfills

Secure landfills are similar to sanitary landfills. However, these landfills make use of additional layers of clay and plastic liners that are designed to prevent LEACHING. They also have expensive monitoring equipment throughout the landfill site to identify a leak if it should occur. In addition, water that collects atop such a landfill is periodically drained and sent to a wastewater treatment plant, where it can be cleaned, purified, and returned safely to the environment.

### Incineration

The incineration process for hazardous wastes is similar to that for residential wastes. However, there are some differences between the two incineration systems. The first involves the temperature at which

◆ Particulates emitted by incinerators are a solid waste problem.

incineration takes place. The second involves the use of special devices such as SCRUBBERS in smokestacks. Such devices are designed to minimize the harmful emissions given off by the burning process. As with other incinerators, disposal of ash remaining in an incinerator used for hazardous waste presents problems. Ashes from hazardous waste incineration are likely to contain harmful substances that can threaten the health of organisms and the environment.

## Deep-well Injection

In some industries, wastes are disposed of through deep-well injection. In this process, wells are dug deep into the earth. The wells are located in rock layers that are below and completely isolated from all groundwater supplies. Once a well is dug, wastes are pumped down into the well. This method of disposal is most commonly used by the PETROLEUM industry for the disposal of unwanted materials pumped out of the ground during the process of OIL DRILLING. Deep-well injection is limited in use because it is difficult to find suitable disposal sites.

## Resource Recovery and Waste Exchange

Resource recovery and waste exchange are fairly recent methods for disposal of hazardous wastes. Materials that are waste to one company may actually be useful to another company. In resource recovery, companies try to recover waste products that may be useful to people working in another indus-

try. These wastes are then sold to the other industry for use in the manufacture of its products. This method of recovering and reusing wastes has benefits to both companies. It reduces the total amount of waste the first company must dispose of and reduces the amount of raw materials the second company must process to obtain a substance.

## INDIVIDUAL RESPONSIBILITY

As people have become more aware of how their actions affect the environment, they have also begun to look for ways to make their activities more environmentally friendly. In addition to seeking out new methods of waste disposal, many people have become more involved in finding ways to reduce the amounts of wastes they generate and throw away. Three ways in which individuals have become more involved in waste disposal concerns are through the practices of recycling, reducing, and reusing.

Recycling involves the reprocessing of discarded materials to make new products. Through recycling efforts, many materials that once made their way to landfills are now made into new products. Currently, materials that are commonly recycled include: paper, glass, PLASTIC, motor oil, and metals such as ALUMINUM, zinc, LEAD, COPPER, silver, iron, and steel.

Many people are trying to reduce the amount of garbage they produce. To aid in this effort, people are beginning to buy products that are intended to be reused instead of buying disposable products. For example, many new par-

ents are using cloth diapers instead of disposable diapers for their children. In addition, companies are trying to use less packaging materials for their products. Both of these efforts cut back on the amount of garbage produced by people.

In addition to recycling and reducing, many people have begun reusing materials that once would have been thrown away. For example, people who have outgrown their clothes may give the clothes to another person instead of throwing them away. In addition to lessening the amount of garbage that must be disposed of, recycling conserves energy and NATURAL RESOURCES and reduces pollution. [*See also* BUREAU OF RECLAMATION; CONSERVATION; CONTAINER DEPOSIT LEGISLATION; ENVIRONMENTAL EDUCATION; HAZARDOUS SUBSTANCES ACT; HAZARDOUS WASTE MANAGEMENT; INDUSTRIAL WASTE TREATMENT; INTERNATIONAL TRADE IN TOXIC WASTE; LABELING, ENVIRONMENT; MARINE POLLUTION; NIMBY; RADIOACTIVE WASTE; RECLAMATION ACT; RECYCLING, REDUCING, REUSING; RESOURCE CONSERVATION AND RECOVERY ACT (RCRA); SEABED DISPOSAL; SOLID WASTE DISPOSAL ACT; SOURCE REDUCTION; SUPERFUND; TOXIC SUBSTANCES CONTROL ACT; and TOXIC WASTE.]

# Waste Reduction

▶Actions and activities intended to reduce the volume and/or toxicity of waste. One key strategy in waste reduction is to bring about changes in consumer behavior through education. Another is to prevent waste "up front" by producing goods that are less wasteful of raw materials.

Some manufacturers have redesigned their products to reduce the amount of packaging required. Packaging requirements are less stringent for concentrates and powders to which water or another liquid is added. Some packaging is itself consumable, such as the dissolvable detergent pack introduced during the 1980s. Technological advances have enabled manufacturers to use less material to make traditional kinds of packages. For example, between the early 1970s and 1990, the weight of a plastic milk jug dropped from 3 ounces to 2 ounces (95 grams to 60 grams).

Waste can be reduced by extending the life of packaging materials through reuse or reconditioning. The use of the once-common refillable milk bottle declined considerably following the introduction of disposable cartons and jugs. It began to reappear in the late 1980s as people became more aware of the waste associated with cartons and jugs.

Bulk packaging can reduce waste, too. Shipping goods, such as fresh produce, in large containers eliminates the relative expense and wastefulness of packing them in small-serving containers.

In addition, manufacturers are making efforts to reduce the use of toxic materials, to increase recovery and recycling, and to increase operating efficiency. Between 1975 and 1993, the Minnesota Mining and Manufacturing Company reduced its overall waste production by one third and emissions of air pollutants by 70% and saved itself more than $600 million in waste-disposal costs. Many manufacturers have begun participating in waste exchanges in which one company's waste is sold to and used by another company. [*See also* CONTAINER DEPOSIT LEGISLATION and RECYCLING, REDUCING, REUSING.]

# Wastewater

▍Fresh water containing a variety of microorganisms and chemicals that are potentially harmful to living things. Wastewater includes the liquid wastes produced by residential areas and industry. It also includes RUNOFF from streets and farmland that enters storm drains. Wastewater carries with it a variety of organic matter and chemicals. If allowed to remain in the water, these substances are pollutants that are harmful to living things. In order to make wastewater safe for use by organisms, it must be sent by pipelines to wastewater treatment plants before it is released back into the ENVIRONMENT.

Wastewater carries a variety of pollutants. Among the more common pollutants are PATHOGENS and the organic matter present in SEWAGE. These pollutants come from POINT SOURCES such as homes, apartment buildings, retail stores, and industries. Other point source pollutants in wastewater may include chemicals such as acids, salts, and toxic metals that are produced by industry or that result from improper disposal of household products such as cleaning fluids, solvents, paints, and DETERGENTS.

◆ Some industries build lined holding ponds for water pollutants to reduce the chance of leakage into the environment.

 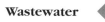

Many of the chemical substances present in wastewater come from NONPOINT SOURCES. An example is any substance carried in runoff that enters water through storm drains. Pollutants that enter wastewater in this way include PESTICIDES, SEDIMENTS, and fertilizers that run off home lawns and gardens as well as farmland. PETROLEUM products such as oil and gasoline that leak onto roadways from cars and salts that are used to melt ice on roadways in winter are other common wastewater pollutants.

Before wastewater is safe for use by living things, it must be sent to a wastewater treatment plant. Such plants remove many types of wastes from water. Clean water is then returned to the environment or sent to homes and industries for use by people. Although wastewater treatment plants are able to filter out many substances from wastewater, they may not be able to remove some types of pollutants. [*See also* SEWAGE TREATMENT PLANT and WASTEWATER, PRIMARY, SECONDARY, AND TERTIARY TREATMENT OF.]

# Wastewater, Primary, Secondary, and Tertiary Treatment of

Different levels of treatment for WASTEWATER that takes place within a wastewater treatment plant. In many areas, laws require that all wastewater, such as industrial wastewater from factories and SEWAGE from homes and businesses, be treated, or cleansed, of harmful materials before being released back into the ENVIRONMENT. Primary, secondary, and tertiary treatment refers to the stages in the process of water treatment that help clean polluting substances from wastewater.

Most wastewater from homes, businesses, and industrial sites is collected in sewer pipes and sent to a wastewater or a SEWAGE TREATMENT PLANT. Here, it is treated before it is released into a lake, river, or other body of water. To remove most of the harmful substances in wastewater, such as organic wastes, heavy metals, PHOSPHATES, and PESTICIDES, a three-step cleansing process is often performed.

## PRIMARY TREATMENT

The first stage of wastewater cleansing is primary treatment. This step of the process is designed to remove large, solid particles, such as leaves and sand, that often float within wastewater. The simplest type of primary treatment system is the cesspool, a filtering system common in rural areas.

A cesspool is a big, underground storage tank that collects wastewater and filters out SOLID WASTE particles. Small holes in the sides and bottom of the tank allow liquid wastewater to flow into the ground while holding the larger particles of solid waste.

In large communities that have wastewater treatment plants, a similar process is performed. Sewer water is passed through a large screen that filters out the solid particles. It is then sent to a grit chamber, which filters out even smaller particles. Finally, after going through both the screen and grit chamber, wastewater is trickled into a sedimentation tank, or settling basin. Here, wastewater is slowly circulated while the remaining solid particles settle to the bottom of the tank. Exit holes near the top of the tank allow the cleansed water to flow out for further processing.

**Wastewater Treatment**

| Levels of treatment | Impurities treated | Used in this type of treatment |
|---|---|---|
| **Primary** | Solid waste, grit, sand, and other large particles | Filtration, sedimentation, suspension |
| **Secondary** | Organic waste and bacteria | Filtration, bacterial decomposition |
| **Tertiary** | Phosphorus and nitrogen compounds, industrial pollutants | Bacterial denitrification and other biological processes, chlorination, radiation, filtration, ion removal or exchange |

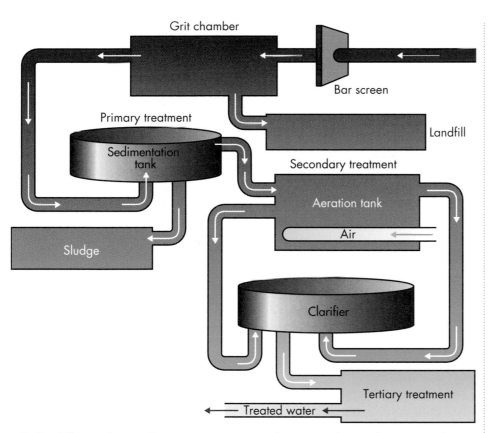

Grit chamber

Bar screen

Primary treatment

Sedimentation tank

Landfill

Secondary treatment

Aeration tank

Air

Sludge

Clarifier

Tertiary treatment

Treated water

◆ In the different phases of wastewater treatment, large particles are first screened out, then organic wastes are removed, and finally, specific pollutants such as phosphates are removed.

Primary treatment removes about half of the suspended particles in wastewater. About 30% of the organic wastes are also removed. Sometimes, chemicals are added to the sedimentation tank. The chemicals cause suspended particles to form clumps that can be easily separated from wastewater. The addition of chemicals makes it possible to remove 80% to 90% of the solid particles in wastewater.

## SECONDARY TREATMENT

Most towns and cities require that wastewater be put through a secondary treatment process. The second process is needed because

primary treatment cannot remove most of the organic wastes in wastewater. Many organic wastes are harmful to the environment because they add too many nutrients to an ECOSYSTEM. Excess nutrients allow populations of ALGAE, BACTERIA, and other microorganisms to flourish, causing population explosions that can rob water of DISSOLVED OXYGEN. The process of secondary treatment involves the DECOMPOSITION of organic wastes in wastewater.

In one type of secondary treatment, known as the *trickling-filter method,* wastewater is slowly sprayed over a layer of gravel and crushed rocks. This slow trickling of

water increases the amount of dissolved oxygen in the wastewater, allowing microorganisms to thrive. Microorganisms covering the rocks break down, or decompose, organic wastes into smaller particles. These particles are then filtered out in a secondary sedimentation tank. The result is SLUDGE, a thick mass of dead bacteria, organic matter, and solid wastes. Sludge is usually dumped into LANDFILLS, burned, or dried in sludge-drying beds for use as fertilizer. Secondary treatment removes about 95% of bacteria and about 90% of organic matter from wastewater.

## TERTIARY TREATMENT

Primary and secondary treatment methods remove most of the harmful substances in wastewater. However, to make wastewater even safer for discharge into waterways, tertiary treatment methods are often used. Tertiary treatment usually involves treating sewage with chemical processes, such as CHLORINATION or RADIATION. Sewage may also be passed through charcoal filters that trap any remaining impurities. [*See also* ALGAL BLOOM; BIOACCUMULATION; CLEAN WATER ACT; DECOMPOSER; EUTROPHICATION; INDUSTRIAL WASTE TREATMENT; MARINE POLLUTION; SAFE DRINKING WATER ACT; WATER POLLUTION; and WATER QUALITY STANDARDS.]

# Wastewater Treatment Plant

*See* SEWAGE TREATMENT PLANT

# Water, Drinking

◗ Fresh water that is free of pollutants and microorganisms and is therefore potable, or fit for drinking by people. Like all other organisms, humans need water to survive. This water is obtained from the food we eat and from freshwater bodies such as lakes, rivers, and groundwater. Water that is potable contains low levels of salts; tastes and smells good; and will not harm the health of those who consume it. In determining water quality, experts consider such factors as water temperature and color, odor, concentration of BACTERIA, DISSOLVED OXYGEN content, and the amounts of elements or chemical compounds the water contains.

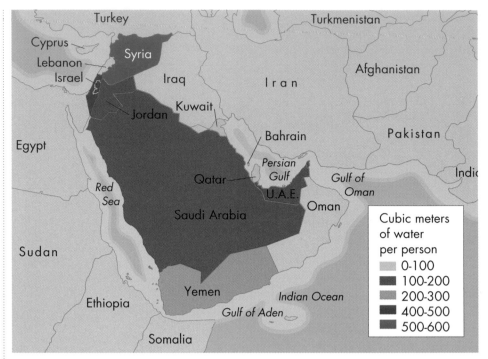

◆ Resulting from a combination of factors, severe water shortages exist in the Middle East.

### HOW MUCH DO WE NEED?

Water is the substance most needed for life. People can survive for weeks without food but only a few days without water. Each day, the average person must drink about 1/2 gallon (2 liters) of water.

### FRESH WATER SUPPLIES

Most of the water on Earth is not fresh water available for use as drinking water. About 97% of Earth's water is salt water. Of the remaining 3% of Earth's water, less than 1% exists in the ATMOSPHERE or in rivers, lakes, or underground AQUIFERS that are accessible to people. While only 1% of Earth's water is available to people, there is enough fresh water on Earth for each person to use.

Water is a renewable NATURAL RESOURCE. The fresh water available to use is constantly cycled through the ENVIRONMENT by the WATER CYCLE. Thus, the water available to people cannot be depleted unless we use it faster than it can be replaced by natural processes or if it is contaminated.

Although Earth has plenty of fresh water for its people, the problem with getting fresh water is distribution. For example, there is plenty of fresh water available in Washington State. However, parts of Arizona, which is largely DESERT, do not have adequate water supplies to meet their needs. These areas obtain the water they need by pumping it in from the Colorado River system.

Another problem is that drinking water in many areas is so polluted or contaminated that it is not fit for human use. The World Health Organization (WHO) estimates that 80% of the disease present in poor, developing countries results from a lack of sufficient clean water. People may get sick or even die from a form of diarrhea that is caused by contaminated water. A disease called hepatitis can spread among people who live where the water is polluted.

### CONTROLLING WATER QUALITY

The only water that is completely free from microorganisms and chemicals is distilled water. Distilled water is pure water that is prepared in a laboratory. However, to most people, distilled water has a strange taste. People are used to drinking water that has been treated with and flavored by some chemicals. The water most people use for drinking, bathing, and cooking also contains very small amounts of microorganisms.

Despite the small amounts of chemicals and microorganisms contained in drinking water, this water is safe for people. The levels of organisms and chemicals that are permitted in drinking water is regulated by agencies such as the U.S. ENVIRONMENTAL PROTECTION AGENCY (EPA). State and local health departments also regulate and monitor water supplies.

One reason people live longer today than they did 200 years ago is because agencies such as the EPA ensure that we have clean water for drinking and for washing ourselves and our dishes. In addition, used and contaminated water called SEWAGE is carried away to SEWAGE TREATMENT PLANTS for processing to keep it from contaminating sources of drinking water.

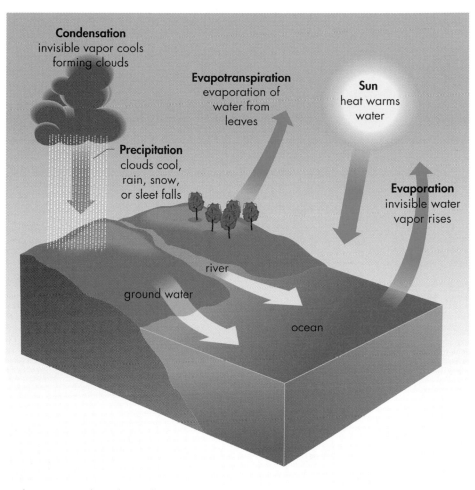

**Condensation**
invisible vapor cools forming clouds

**Evapotranspiration**
evaporation of water from leaves

**Sun**
heat warms water

**Precipitation**
clouds cool, rain, snow, or sleet falls

**Evaporation**
invisible water vapor rises

river

ground water

ocean

◆ The water cycle is the endless movement of water from Earth to the atmosphere and back again.

# Water Cycle

▶The continuous movement of Earth's water from Earth's surface into the ATMOSPHERE and back to the surface. This never-ending recycling and cleaning process is also known as the *hydrologic cycle*.

Because it is a cycle, the process has no true beginning step. However, it is easiest to follow the water cycle from the point at which the sun heats SURFACE WATER in Earth's OCEANS, lakes, streams, and puddles. Some drops of heated ocean water scatter into the air as invisible water vapor, leaving any salts, MINERALS, and other solid particles behind. The process by which liquid water changes to a gas is called *evaporation*. As the vapor rises and cools, water droplets condense around dust particles in the air to form CLOUDS. The cooling of water vapor back into a liquid is called *condensation*. When the clouds become saturated, or unable to hold any more water, PRECIPITA-TION falls back to Earth as rain, snow, sleet, or hail. The type of precipitation that falls depends on the air temperature.

Most precipitation falls directly into oceans, rivers, and streams; the rest falls on land. Some evaporates directly from rooftops and ground puddles. Some runs off the land into streams and rivers, which carry it back to the ocean. The remaining water soaks into the SOIL to become part of Earth's groundwater, from which it slowly moves to rivers that carry it back to the ocean. There, the water cycle begins again.

Because water constantly cycles through the ENVIRONMENT, rain or snow that falls today has fallen billions of times before. The water may have changed form—from liquid, gas, or solid—and may have moved from one place to another, but the amount of water on Earth today is the same as it has been for millions of years. So the water you drink may be the same water in which prehistoric **plesiosaurs** swam, on which ancient explorers sailed, and that George Washington drank.

## CLEANING DURING RECYCLING

Ocean water is too salty for drinking, agriculture, or most industrial processes. But during the water cycle's evaporation process, salt contained in ocean water is left behind. Precipitation falls as fresh water.

You can duplicate this water-cleaning process. Hold a pan filled with ice cubes six inches (15 centimeters) above a pot of boiling salt water. Note the white **crust** (salt) that forms on the edge of the boiling pot. Vapor rising from the salt water condenses on the bottom of the ice-filled pan, then falls back into the pot. Carefully rub your finger across the condensed water on the bottom of the ice-filled pan and taste it. It is fresh, not salty, because the salt was left behind during evaporation.

## THE WATER CYCLE SHAPES EARTH

Water moving through the water cycle changes Earth's surface. For example, precipitation that falls at high elevations is pulled downhill by gravity, eroding soil and rock as it flows to create mountains, canyons, and river channels. Sometimes heavy precipitation makes rivers and streams move with such great force that they change course or overflow, carrying soil and other debris downstream. When the water movement slows, it will drop its cargo, building up Earth's surface. A common feature formed by

---

**THE LANGUAGE OF THE ENVIRONMENT**

**crust** hard, brittle deposit formed around the edges of an object.

**delta** low plain, usually triangular, formed at the mouth of a river by deposits of sand and soil.

**plesiosaurs** large water reptiles of the Mesozoic Era; prehistoric sea creatures.

---

such movement is the **delta**. A delta is a plain shaped like a triangle that forms from SEDIMENTS dropped at the mouth of a river. Precipitation may pile up to become GLACIERS that grind and scrape across Earth's surface, cutting sharp, jagged peaks and dumping rock, dirt, and gravel. And finally, the ocean's waves erode beaches and carve out new shorelines on Earth's landforms.

## THE WATER CYCLE AND THE ENVIRONMENT

Usually, Earth's water cycle supplies clean water for drinking, farming, and industry. But sometimes SULFUR DIOXIDE and other pollutants from factories and cities mix with the water vapor in the air. The result is ACID RAIN, chemically strong enough to erode statues and buildings and poison lakes, FORESTS, and farmland. Acid rain has "killed" some lakes and rivers in the United States and Canada, making them incapable of supporting PLANT or animal life.

Recycled water is a RENEWABLE RESOURCE and one of our ALTERNATIVE ENERGY SOURCES. For example, DAMS use moving water to turn giant turbines that produce ELECTRICITY. Other industries use the steam from boiling water to power machinery.

When people use water, we sometimes pollute it. We try to clean polluted water through water-purification processes, such as filtering or adding chemicals such as chlorine to kill BACTERIA. However, some pollutants cannot be removed easily. For example, TOXIC WASTES that are not disposed of safely can leak chemicals into underground water sources. This often poisons drinking water that fell as clean precipitation.

The 1972 U.S. CLEAN WATER ACT, and subsequent revisions, limited pollution of our water and has resulted in much cleaner water supplies in the United States. [*See also* AQUIFER; DESALINIZATION; EVAPOTRANSPIRATION; GEOTHERMAL ENERGY; LEACHING; MARINE POLLUTION; OXYGEN CYCLE; RUNOFF; SAFE DRINKING WATER ACT; and WATER QUALITY STANDARDS.]

# Water Pollution

❚Contamination of water by natural events, by animal wastes, and by wastes produced by humans and by their activities. Water pollution has been a serious problem affecting aquatic ECOSYSTEMS of many kinds. Water pollutants can adversely affect WILDLIFE and humans. Fresh water covers a very small part of the Earth's surface (about 3%). Humans and many other SPECIES are dependent on this water supply. Salt water covers about 70% of Earth. As the OCEANS become polluted, the task of cleaning up this water becomes a worldwide concern. In the United States, water pollution is regulated primarily by the CLEAN WATER ACT.

Fresh water is produced constantly as a result of the WATER CYCLE. Fresh water is present in many kinds of ecosystems (streams, rivers, lakes, puddles, and many kinds of WETLANDS), and in groundwater. Humans use water from these sources in many ways: for drinking, for IRRIGATION, for cooling in industrial processes, and for creating ELECTRICITY in HYDROELECTRIC POWER plants. Human populations also expect to use water ecosystems for FISHING, swimming, and other forms of recreation.

## POLLUTION SOURCES

Water pollution comes from different kinds of sources. Sources of water pollution that come from specific locations such as pipes or ditches are referred to as POINT

SOURCES. NONPOINT SOURCES are scattered areas from which pollutants enter SURFACE WATER and groundwater. These areas include city streets and parking lots that contaminate storm water and feedlots, farmland, and lawns from which come fertilizers, PESTICIDES, and eroded SOILS. Air pollutants that produce ACID RAIN can also be thought of as nonpoint sources. Because point sources of pollution are easy to identify, they are easier to control and regulate than nonpoint sources. The majority of pollutants that affect water come from nonpoint sources, but pollutants from both point and nonpoint sources continue to affect aquatic ecosystems.

## TYPES AND EFFECTS OF WATER POLLUTION

SEWAGE and animal wastes can carry into surface waters such organisms as BACTERIA and protozoans, and other PATHOGENS, such as VIRUSES, that can cause disease. These diseases affect people who have contact with contaminated water or who drink untreated water. Diseases can also be transmitted by eating FISH or shellfish caught in contaminated water.

### Organic Pollutants

Some wastes such as sewage and those formed by food processing

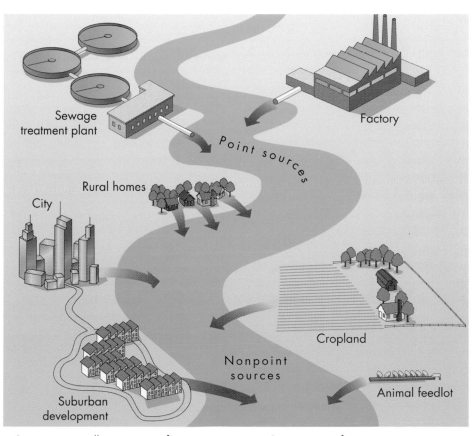

Sewage treatment plant

Factory

Point sources

City

Rural homes

Cropland

Nonpoint sources

Suburban development

Animal feedlot

◆ Some water pollution comes from point sources. Some comes from nonpoint sources, such as city street runoff.

can be broken down by DECOMPOSERS in water. As these wastes are broken down, large amounts of OXYGEN may be used up by the decomposers. As a result, oxygen is removed from the water. A drop in the dissolved OXYGEN level may cause many organisms that need high levels of oxygen to die. Even when oxygen is not completely removed, this kind of pollution can restrict the types of species that can survive in the water. Wastes of this kind are known as oxygen-demanding wastes.

Some nitrogen and phosphorus compounds can cause excess growth of ALGAE, a direct result of accelerated EUTROPHICATION. When large numbers of algae die, their decaying remains act like oxygen-demanding waste. Some forms of nitrogen, like nitrates and ammonia, are toxic to humans and other organisms. In fresh water, excess phosphorus in the form of PHOSPHATES produces excess algae growth called ALGAL BLOOMS. This kind of pollution can come from point sources like SEWAGE TREATMENT PLANTS and from nonpoint sources such as fertilizers and soils eroded from farms, FORESTS, and city lawns.

Most commercially produced chemicals are organic. Organic chemicals are used in many industrial processes, ranging from the manufacture of PLASTICS to the production of other chemicals, including pesticides. Additionally, cleaning solvents and DETERGENTS can be water pollutants. Oil, gasoline, and other PETROLEUM products can cause a drop in the oxygen level in water. These products also have components that are toxic to many organisms.

## Inorganic Pollutants

The most important inorganic pollutants are toxic metals such as arsenic, cadmium, COPPER, chromium, LEAD, MERCURY, selenium, and zinc. Some inorganic pollutants, such as acids from ACID RAIN, can dissolve other toxic metals like ALUMINUM to pollute streams and lakes.

Radioactive substances are not commonly disposed of in water. However, when they are, they can enter FOOD CHAINS and cause mutations, genetic damage, and CANCER.

## Natural Pollutants

Particles of soil, silt and other suspended solids mixed in fresh water are known as *suspended particles*. These particles create a sig-nificant water pollution problem. Suspended particles cloud the water and settle to the bottom of streams, rivers, and lakes, covering the water bottom with mud and muck that smothers organisms living there.

The soot and ash from volcanic eruptions are also a source of particles that pollute water. These and other suspended particles reduce PHOTOSYNTHESIS by preventing light from penetrating in the water. They also make HUNTING difficult for species that need to see their prey. Some particles often have toxic materials like pesticides or inorganic chemicals attached to them, as well as bacteria, nutrients, and toxic metals. The smothering effect of particles fills RESERVOIRS and

◆ Ocean pollution is a worldwide concern because pollution from one country may be carried by currents to another country.

destroys feeding and breeding areas for fish.

## Thermal Water Pollution

THERMAL WATER POLLUTION occurs when water is heated. Heated surface water or groundwater that has been used for cooling water in industrial processes is often disposed of in lakes and streams. Warm water holds less oxygen than cool water. Changing the temperature of surface water can affect the rate of development of fish eggs. It can also affect many species through heat stress or by promoting disease.

## WATER POLLUTION CONTROL

Advances in the control of water pollution have been made on many fronts, but there is still much work to do. The Clean Water Act established national goals for WASTEWATER treatment and prohibited the release of toxic chemicals in dangerous amounts in the waters of the United States. Regulation of toxic chemicals also occurs under several other laws such as the TOXIC SUBSTANCES CONTROL ACT and the FEDERAL INSECTICIDE, FUNGICIDE, AND RODENTICIDE ACT (FIFRA).

The Clean Water Act led to the creation of national WATER QUALITY STANDARDS for 129 pollutants and the development of methods for measuring the level of toxicity. The act has also helped communities build sewage treatment plants and has required cities, towns, and industries to control waste discharges under the NATIONAL POLLUTANT DISCHARGE ELIMINATION SYSTEM (NPDES).

Control of point sources has been effective, but the control of nonpoint sources is a more complex problem. For example, the United States and Canada have signed agreements to eliminate the discharge of over 300 toxic chemicals into the GREAT LAKES, but studies have shown that 50% of the input of toxic chemicals into Lake Superior, the largest of the Great Lakes, are from the ATMOSPHERE, a nonpoint source. About two-thirds of the contamination of surface waters may be from nonpoint sources. In a way, nonpoint sources are everyone's problem, since water running off streets, lawns, farm fields, and forests can add to nonpoint pollution. This means that many people will have to help in solving the nonpoint pollution problem. [*See also* ARTESIAN WELLS; BIOCHEMICAL OXYGEN DEMAND (BOD); CHLORINATION; DAMS; DREDGING; ESTUARY; HYDROLOGY; HYDROSPHERE; INFILTRATION; LEACHING; MARINE POLLUTION; MEDICAL WASTE; MINAMATA DISEASE; OCEAN DUMPING; OIL SPILLS; RESERVOIR; RUNOFF; SAFE DRINKING WATER ACT; SALTWATER INTRUSION; SEDIMENT; and SEDIMENTATION.]

# Water Purification
*See* WATER QUALITY STANDARDS

# Water Quality Standards

▯Federal limitations on the amount of contamination allowed to enter SURFACE WATER—rivers, lakes, and streams. The CLEAN WATER ACT of 1972 set a national goal of "fishable, swimmable" surface waters and permitted each state to set its own water quality standards. These standards are subject to approval by the ENVIRONMENTAL PROTECTION AGENCY (EPA).

Supervision of water quality is difficult because no single federal organization is in charge. It is a joint responsibility of the EPA, the U.S. FISH AND WILDLIFE SERVICE, the NATIONAL OCEANIC AND ATMOSPHERIC ADMINISTRATION (NOAA), and the U.S. Geological Survey.

## SETTING NATIONAL STANDARDS

Following conditions in the Clean Water Act, each state presented a written plan to the EPA listing exact amounts of common pollutants, such as acids, **fecal coliform** BACTERIA, and oil and sediments, it considered acceptable in its water. The state also had to give the guideline it used to set its standard—usually scientific data about the amounts of organisms or a substance that constitutes toxic levels.

The Clean Water Act covered both water quality and **technology-based standards** for surface waters. Technology-based standards required community SEWAGE plants to use a second treatment to help purify water before releasing it into rivers. The Clean Water Act originally required all industries discharging WASTEWATER into rivers to use "Best Practical Technology" (BPT) by 1977 and the "Best Available Control Technology (BACT) Economically Achievable"

by 1983. The 1987 Water Quality Act used water quality rather than technology-based standards, saying the cost or added bother of using the best technologies to meet standards was no excuse for letting water fall below standard.

In 1974, Congress passed the SAFE DRINKING WATER ACT. The act authorized the EPA to set quality standards for public water systems to reduce amounts of harmful bacteria, chemicals, and metals in drinking water. In 1979, the EPA added limitations on the allowable amounts of chloroform and other chemicals known as *trihalomethanes* (THM) in the drinking water. These chemicals formed when chlorine was added to water at treatment plants to kill disease-causing bacteria. It was believed that exposure to high levels of these chemicals might increase the risk of CANCER.

Water quality and technology-based standards for EFFLUENT from sewage plants into surface water are set through a permit system. The state or EPA issues POLLUTION PERMITS under the NATIONAL POLLUTANT DISCHARGE ELIMINATION SYSTEM (NPDES).

## INDUSTRY AND WATER

Industry uses 52% of all water used in the United States. However, only 2% goes into products. Most of the water is used for cooling processes, then discharged into the river or lake from which it came.

Almost every industrial activity results in some form of waste—from harmless sand to toxic material. Various sewage treatments remove large amounts of toxic chemicals and heavy metals from industrial waste, but 10% to 20% may be discharged into surface waters untreated. The Toxic Release Inventory (TRI), set up in 1987 to monitor environmental discharges from U.S. industry, reported in 1989 that manufacturers released 189 million pounds (85 million kilograms) of toxic material into streams, lakes, and rivers.

## FUTURE STANDARDS

The Clean Water Act limits the pollutants industry can dump into surface waters, requires small sewage plants to update their methods,

◆ Laboratories worldwide must test local water samples to see if the water is polluted or not.

limits destruction of woodlands, and requires communities to treat polluted stormwater. Some groups would like to revise the act to increase industrial pollutants allowable for dumping; postpone for 15 years the updating of small sewage plants; leave up to 80% of WETLANDS unprotected; and give states control over stormwater pollution. Environmentalists continue to fight against changes that they claim would undo much of the progress that has been made thus far. They point out that stormwater adds up to 34% of the LEAD, 21% of the COPPER, 14% of the MERCURY, and 15% of the polychlorinated biphenyls (PCBS) that end up in rivers. Wetlands provide food and nurseries for FISH and serve as pollution filters. [*See also* BIODEGRADABLE; HEAVY METALS POISONING; LEACHING; MINING; PATHOGEN; POLLUTION PERMIT; WASTEWATER, PRIMARY, SECONDARY, AND TERTIARY TREATMENT OF; and WATER TREATMENT.]

# Water Rights

❚A legal right to use fresh water from a particular source. Fresh water includes groundwater, rivers, streams, lakes, and ponds. Since fresh water is a scarce NATURAL RESOURCE, most societies have developed laws that define how much water a person or company can take from a source.

In areas where water is in short supply, arguments about water rights can be intense. In the western United States, for example, farmers and urban dwellers are sometimes in conflict. Most fresh water is used to irrigate farmland, and water rights are often sold to farmers for lower prices. Cities often argue that it would be more economical to use the same water for household and industrial use, because the water rights could be sold for much higher prices.

When a river flows from one country to another, the situation becomes even more complicated. All the countries affected must agree on how the water will be distributed, or conflicts may result. [*See also* IRRIGATION; WATER, DRINKING; and WATERSHED.]

# Watershed

❚The area of land that is drained by a river, stream, or lake. When rain, sleet, snow and other forms of PRECIPITATION fall over land, some of the water soaks into the ground. This water is called *groundwater*. The remaining water, called SURFACE WATER, runs down mountains and hills and across flat plains as small streams. These connect to form larger streams and eventually join to form rivers. The watershed is the total area of land drained by these streams.

Depending on the size of a river, an area's watershed can be

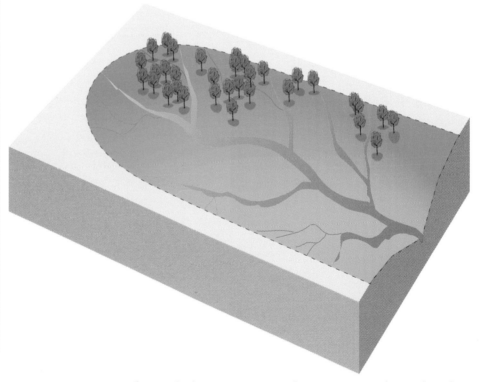

◆ An important aspect of watershed management involves maintaining the quality of soils in the watershed area. Clear-cutting forests on slopes increases erosion of soils, which can disrupt a watershed.

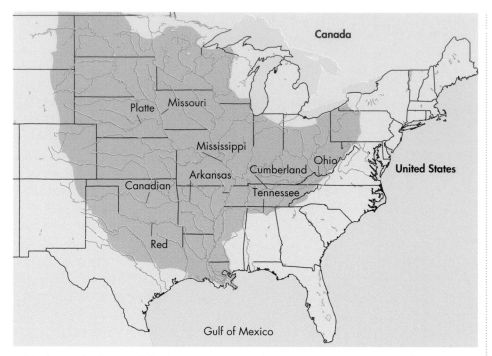

◆ The watershed area of the Mississippi River, the longest river in the United States, covers much of the central portion of the country.

quite large. The watershed of the central portion of the Mississippi River, for example, covers most of the central portion of the United States. In most undisturbed watersheds, much of the water moves slowly through the ground. It is sometimes delayed in ponds and lakes, allowing a variety of living things to use the water and helping to maintain the quality and purity of the water.

Natural, undisturbed watersheds have clear streams and an ample covering of trees, grasses, and other PLANTS. When precipitation falls, some of the water is scattered over leaves and branches. Much of this water returns to the ATMOSPHERE through evaporation. The remaining water soaks into the ground through tiny channels within the SOIL. Here, it is absorbed by the roots of plants and used for PHOTOSYNTHESIS.

In healthy ECOSYSTEMS, TOPSOIL is filled with countless holes and spaces that can absorb great amounts of water. The level at which the ground is permanently saturated with water is known as the WATER TABLE.

The amount of water in an area's water table changes with the season and is related to the amount of precipitation the area receives. During periods of heavy rain, the ground may not be able to soak up all the water. Much of it flows along the surface in streams.

## WATERSHED MANAGEMENT

When FORESTS are cleared for agriculture or when grasses are stripped by GRAZING animals, an area's watershed can be disturbed. These activities decrease the quality of soils by depleting nutrients and increasing EROSION.

When there are no trees or grasses to break the force of falling rain, mud clogs the tiny channels in soil, which slows the soaking of water into the ground. If the land is flat, water collects in stagnant pools. In sloped areas, surface waters can race downhill, uprooting grasses and other small plants along the way. Fast-moving water carries away soil.

Disturbing watersheds increases the chances of flooding in an area. In addition, the level of water in the water table is also changed. Water from heavy rains or melting snow is often prevented from fully soaking into the ground. Such water can overflow a riverbank, causing flooding in the spring. In the summer, streams and AQUIFERS can dry up because little or no water has sunk into underground RESERVOIRS. Effective watershed management, therefore, must include techniques to protect the health of ECOSYSTEMS and maintain the quality of soils. [*See also* CLEAR-CUTTING; FOREST MANAGEMENT; SEDIMENTATION; SOIL CONSERVATION; WATER CYCLE; and WATER POLLUTION.]

# Water Table

❯The upper surface of the water-bearing layer of Earth's crust. The WATER TABLE lies at the upper surface of an AQUIFER. An aquifer fills as rain

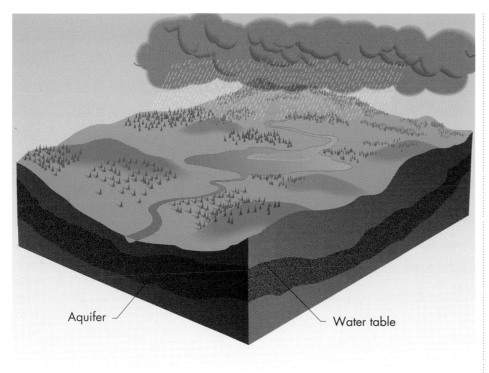

◆ Seasonal rainfall affects the level of a water table.

percolates down through SOIL into layers of rock and gravel. Some of the spaces between rock and gravel contain water, while others contain air. This area of an aquifer is the zone of aeration. Beneath the zone of aeration is a rock layer in which all spaces are filled with water. This area of an aquifer is the ZONE OF SATURATION. The water table is the upper surface of water contained in the zone of saturation.

Several events can raise or lower a water table. The amount of rainfall an area receives is one of these factors. In addition, irrigating farmland often raises the water table, in some cases making the soil waterlogged and useless. In many areas, water is pumped from aquifers for use by people. Such activity can lower the water table. If the water table becomes too low because water is removed faster than it is replaced, it becomes more expensive to pump water out of the aquifer. In addition, if an aquifer is depleted of water, the ground may subside. [*See also* ARTESIAN WELLS; LEACHING; SALTWATER INTRUSION; and WATER, DRINKING.]

# Water Treatment

▮Preparation of water to make it safe for human consumption. Water is absolutely vital to life, and yet it may be full of dangerous impurities and PATHOGENS, microscopic organisms that cause disease. Before water treatment became standard practice in the United States and other developed nations, typhoid fever was common, and outbreaks of cholera devastated whole communities. Such diseases, sometimes called sanitation-related diseases, are caused by drinking water that has been contaminated as a result of lack of proper sanitation. Safe drinking water is still difficult for people to obtain in many developing nations and even in some developed nations. Poland, for example, has notoriously bad drinking water.

Communities in the United States have adopted various methods of removing organisms and impurities from water. Some cities use a combination of methods to purify their water. In many large cities in the United States, the first, or primary, stage of water treatment is long-time storage in RESERVOIRS; in some communities, in fact, it is the chief method of water treatment, supplemented by the addition of chemicals. Storage reduces the velocity of water, causing particles suspended in it to settle to the bottom. The rate at which particles settle depends on their size, weight, and shape. Water treatment plants sometimes use large sedimentation basins to the same effect.

Storing water also reduces nitrates and BACTERIA. When water is taken from a reservoir or river, it passes through intake valves that are equipped with screens to keep out FISH, PLANTS, and trash. Since the end of World War II, some systems have been equipped with microstrainers on revolving drums with 0.001-inch openings to screen out very small impurities.

◆ The larger the population served, the bigger the water treatment facility must be.

Screening does not remove all impurities, however. The water may next be pumped to a filtration plant and allowed to trickle down through layers of sand and gravel. This filters out larger particles of debris and grit. From the filtration plant, the water may go to a chemical station. Here, chlorine is used in small quantities to kill bacteria. Lime and alum are also added in measured amounts to form a colloid of the compound aluminum hydroxide. This fluffy, gel-like colloid forms flakes or globs of sticky gelatin, called *floos*, which collect bacteria and other finely dispersed impurities that escaped filtering. After being rapidly and thoroughly mixed, the mixture may be filtered through beds of fine sand or passed very slowly through a sedimentation basin or tank. Because the water in the basin or tank is barely moving, floos and the impurities they have collected settle as SEDI-MENTS to the bottom, from which they can be removed later. The water is again filtered through layers of sand, gravel, and hard coal. Bad odors can be removed either by the activated carbon, which absorbs gases, or by aerating the water—spraying it into the air to add OXYGEN—after it leaves the filtered water reservoir. More chlorine may then be added to control bacterial growth.

# Weather

▌Conditions of the ATMOSPHERE in a particular place at a given time. Weather can change from one hour to another. However, over many years, some conditions become characteristic of an area. These average day-to-day conditions, including variations and rarities, determine an area's CLIMATE.

## THE IMPORTANCE OF WEATHER

Weather has always been important to humans because it affects everyone's life. People who make a living from the sea need to know if a storm will threaten their fishing boats. Farmers need to know if it will rain or if storms might ruin their harvesting activities. In addition, all people need time to prepare for storms that might damage homes or take lives.

Today, weather forecasts still help humans plan activities. For example, power plants can prepare for increased energy use during winter storms or summer heat waves. Ships and aircraft can plot routes to avoid bad weather. Construction companies can plan around delays for extreme cold, heavy rain, or high wind. In addition, people can plan what to wear each day and also plan for outdoor activities such as boating, ice skating, golf, skiing, cookouts, ball games, and other excursions based on a knowledge of the weather.

## WHAT MAKES WEATHER?

Weather is determined by four components: temperature, air pressure, wind, and moisture. Most weather starts in the TROPOSPHERE, the layer of air closest to Earth. Because of the way Earth rotates on its axis, weather in the northern hemisphere generally moves from west to east.

SUNDAY, JANUARY 7, 1996

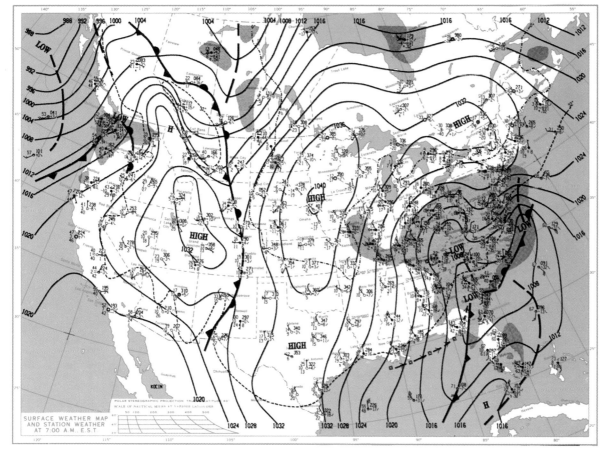

◆ Weather maps are marked to show high- and low-pressure areas, cloud cover, precipitation, wind speed, and more. These same symbols are used worldwide to prevent problems in communication.

Thermometers measure the temperature, or degree of heat in the air, on the Celsius (C) and Fahrenheit (F) scales. The heat comes from the sun's rays and the reflected warmth from sun-warmed land and water.

Barometers measure air pressure—the weight of the air pushing on Earth—in millibars or inches of MERCURY. Air pressure at sea level is about 29.9 inches (1,013 millibars) of mercury; high pressure is about 30.4 inches (1,030 millibars); and low pressure is about 29.4 inches (995 millibars). Warm air, which weighs less than cool air, forms low-pressure areas; cool air forms high-pressure areas. Changes in air pressure signal weather changes.

Winds blow from areas of high pressure to areas of low pressure. The greater the difference in pressure, the stronger the wind. *Anemometers* are devices used to record the wind's speed and the direction from which it is blowing.

Moisture enters the air as water vapor. Most of this water is changed from a liquid to a gas as the sun heats the OCEAN's water. *Humidity* is the amount of moisture in the air. *Relative humidity* is the percentage of moisture compared to what the air can hold at a given temperature. A relative humidity of 100% means the air is saturated. The temperature at which it becomes saturated is its *dew point.*

When warm, moist air rises and cools, water vapor in the air condenses into water drops that form CLOUDS. If warm, moist air near the ground cools to its dew point, low clouds or *fog* may develop.

*Rain gauges* measure the amount of PRECIPITATION that falls from the clouds. Precipitation is rain, snow, sleet, or hail, and is dependent on the temperature. Moisture from clouds falls as rain. However, if air near the ground is below 37° F (2.78° C), the rain turns to snow. At about 39° F (3.89° C),

**short-range and extended forecasts** predictions of weather conditions for a period of time. Short-range forecasts predict weather for the next 18 to 36 hours; extended forecasts predict weather for 5 to 10 days and also for a 30-day period.

it turns to sleet. Hail forms when strong air currents toss ice crystals up and down within high clouds until their mass becomes great enough to cause them to fall from the clouds.

Sometimes air pollutants such as SULFUR DIOXIDE mix with water drops in clouds to create ACID RAIN. Acid rain erodes buildings and can poison land and water.

## MONITORING EARTH'S WEATHER

The NATIONAL OCEANIC AND ATMOSPHERIC ADMINISTRATION (NOAA), a division of the U.S. Department of Commerce, is charged with regulating the use of, and directing scientific research about, our oceans and atmosphere. The NATIONAL WEATHER SERVICE, part of NOAA, monitors weather around the world, reports current conditions, makes **short-range and extended forecasts**, and alerts the public to possible NATURAL DISASTERS such as hurricanes and floods.

Local weather forecasters, usually trained as meteorologists, use data from the National Weather Service, including satellite photos and reports from weather ships and planes, to help forecast weather for an area. They also use local instruments that record conditions at ground level, data from weather balloons, and computer models. Even with all this information, meteorologists cannot always accurately predict the weather. A slight variation in temperature, air pressure, wind, or moisture can suddenly change conditions. [*See also* BIOME; CLIMATE CHANGE; EL NIÑO; GLOBAL WARMING; GREENHOUSE EFFECT; METEOROLOGY; NUCLEAR WINTER; OZONE HOLE; OZONE LAYER; and WATER CYCLE.]

# Weathering

▯Physical and chemical processes by which rocks are broken down. There are two kinds of weathering—mechanical weathering and chemical weathering. Together, these two processes break apart rock and change its composition.

In mechanical weathering, rocks are broken apart and changed in size and shape. Such changes result from natural forces such as the movements of burrowing animals, pressure exerted by PLANTS' roots, heating and cooling, freezing and thawing, and wetting and drying.

In chemical weathering, the MINERALS and compounds that make up rocks are changed chemically. This action results in the formation of new compounds. The natural activities that result in chemical weathering involve the action of OXYGEN, water, and other chemicals such as acids, carried by wind and rain.

The rock dust that is produced by weathering is called *regolith*. Through continued weathering and DECOMPOSITION, SOIL develops from regolith. [*See also* ACID RAIN and EROSION.]

# Wetlands

▯HABITATS, such as marshes, swamps, and bogs, that are saturated by water for at least part of the year. In general, wetlands are transitional areas between terrestrial and aquatic ENVIRONMENTS. Such transitional areas are called *ecotones*.

Ecologists recognize two general types of wetlands—inland wetlands and coastal wetlands. Within these categories are many specific types of wetlands, including SALT MARSHES, ESTUARIES, PRAIRIE potholes, wet meadows, MANGROVE swamps, and cypress swamps.

In the United States, more than 90% of wetlands are inland. The remaining 10% are coastal wetlands located near the OCEAN. All 50 states have some wetlands. However, the greatest number of wetlands are concentrated in only a few states, including Alaska, Florida, Louisiana, and Minnesota.

## BIODIVERSITY IN WETLANDS

Wetlands teem with life. They are among the most productive and diverse ECOSYSTEMS on Earth.

Wetlands are generally rich in PLANT life. The diversity of plants supports large numbers of animals. In fact, many SPECIES are unique to wetlands. The National Audubon Society estimates that some 5,000 plant species, 190 AMPHIBIAN species, and one-third of all BIRD species in the United States depend on wetlands for at least a part of the year. Inland wetlands, for example, are home to many species of ducks, geese, songbirds, freshwater FISH, lizards, frogs, salamanders, and MAMMALS. Coastal wetlands provide HABITATS for marine fish, shellfish, snails, and a variety of birds and mammals. Both types of wetlands are also home to millions of microscopic creatures, including many species of FUNGI, BACTERIA, protists, and ALGAE. Many of these organisms are DECOMPOSERS, which feed on dead and decaying plants and animals.

## IMPORTANCE OF WETLANDS

In addition to their importance to WILDLIFE, wetlands are also important to people. Wetlands often prevent flooding of both inland and coastal areas. In addition, more than two-thirds of all saltwater fish, shellfish, and some freshwater

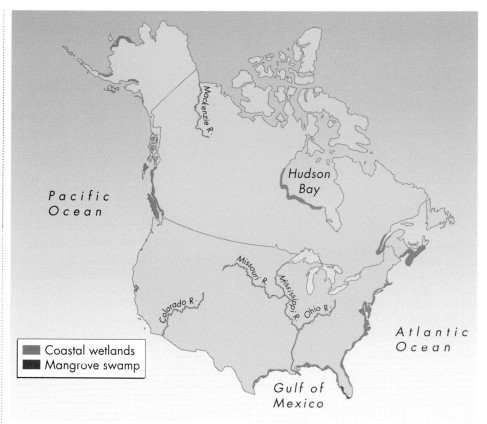

Coastal wetlands
Mangrove swamp

◆ Today, wetlands are being lost at an alarming rate. Many wetlands are drained, filled in, and used for agriculture.

game fish harvested in the United States for use as food use wetlands at some point in their life cycle.

An important benefit of wetlands is their function as natural water purifiers. Wetlands can improve water quality by trapping and filtering out heavy metals, PESTICIDES, toxic chemicals, and SEDIMENTS in water. Many communities in states such as Florida, Kentucky, Tennessee, Georgia, and California make use of local wetlands to cleanse discharged WASTEWATER.

## LOSS OF WETLANDS

Today, wetlands are being lost at an alarming rate. Human population growth has led to the conversion of wetland areas to farmland, housing developments, and other development projects. Despite renewed efforts to protect them, millions of wetlands are lost each year in the 48 mainland states, mostly for agricultural use. In fact, the U.S. FISH AND WILDLIFE SERVICE has estimated that more than half the original wetlands of the United States have been destroyed since the 1700s. Freshwater wetlands have been hardest hit. Between the mid-1950s and mid-1970s, nearly 6.3 million square miles (16.3 million square kilometers) of these wetlands were drained and converted into farmland.

Estuaries and coastal wetlands are also at risk. Until recently, coastal wetlands were seen as bothersome mosquito-infested wastelands. Many of these areas have been drained and filled in for new housing developments. Others have been used as dumping grounds for waste. In the United States, 55% of all estuaries and coastal wetlands have been damaged or detroyed.

## PROTECTING AND RESTORING WETLANDS

Conservationists fear that further loss of wetlands will result in great losses to Earth's BIODIVERSITY. In North America, for example, the four main waterfowl MIGRATION routes cross important wetlands areas. Ducks, geese, and other migratory birds use wetlands for breeding and resting grounds because of the abundance of food and water in these areas.

The CLEAN WATER ACT protects millions of square miles of wetlands from destruction. However, this portion of the act has become a political battleground in recent years. Many people in Congress and elsewhere favor new laws that redefine wetlands. Critics of such laws argue that the proposed changes would result in a loss of nearly 80% of the wetlands that remain.

However, there are some very successful wetlands protection programs in the United States. For example, since 1986, the North American Waterfowl Management Plan (NAWMP) has protected and restored more than 6 million square miles (15.5 million square kilometers) of North American wetlands.

NAWMP supporters include government representatives from the United States, Canada, and Mexico, as well as private CONSERVATION groups. Under NAWMP, more than $547 million has been invested in wetlands conservation. Under the 1990 North American Wetlands Conservation Act, the federal government is authorized to spend up to $26 million annually for wetlands projects. Much of this money is used to fund NAWMP projects. [*See also* EVERGLADES NATIONAL PARK; FISHING, COMMERCIAL; HABITAT LOSS; RESTORATION BIOLOGY; WATER, DRINKING; and WETLANDS PROTECTION ACT.]

# Wetlands Protection Act

▶A law designed to protect U.S. WETLANDS. A wetland is an area of land covered by water at least part of the year. Wetlands include ESTUARIES, swamps, and marshes.

In the past, wetlands were considered to be waste areas. Much of the wetlands in the United States had been destroyed by the 1970s. Since then, the ecological importance of wetlands has been recognized. Efforts are now being made to preserve the remaining wetlands of the United States.

The Wetlands Protection Act is the short name for the Emergency Wetlands Resources Act of 1986. This act authorizes the U.S. FISH AND WILDLIFE SERVICE to designate and

buy important wetlands. The service is also making an inventory and map of wetlands in the United States. [*See also* EVERGLADES NATIONAL PARK; HABITAT LOSS; and MANGROVES.]

# Whales

▶Marine MAMMALS classified in the order Cetacea, a group of animals that also includes DOLPHINS/PORPOISES. Whales are further classified into two groups based upon the structures that they have for feeding. The two groups of whales are the baleen whales *(Mysticeti)* and the toothed whales *(Odontoceti)*.

## GENERAL TRAITS OF WHALES

Whales are one of the most unusual groups of mammals. Like other mammals, such as dogs, cats, cows, ELEPHANTS, and humans, whales are endothermic, or warm-blooded, give birth to live young, nurse their young on milk, and breathe air. However, unlike most other mammals, whales are completely aquatic.

Many SPECIES of whales form and travel in well-organized social groups. Some species migrate thousands of miles each year in search of food or suitable areas in which to give birth to their young. As they travel, many species navigate using *echolocation*. Echolocation is a means by which sound is bounced

◆ Humpback whales feed in groups off the coast of southeast Alaska.

◆ Baleen is a growth of long plates that hang in the whale's mouth from inside the upper lip.

off an object in order to determine its size, movement, and location. In addition, most whales have developed a means of using sound to communicate with others of their species. Often sound communication involves slapping the surface of the water with tails and fins. However, in some whales, very high-pitched tones (toothed whales) or very low-pitched tones (baleen whales) are made. Whales may use such tones to communicate with others of their species over distances as great as thousands of miles.

## BALEEN WHALES

There are about ten species of baleen whales. Although these whales, which lack teeth, feed mostly on very small FISH, tiny crustaceans (such as **krill**), and microscopic PLANKTON, most baleen whales grow to be quite large. In fact, the endangered blue whale,

the largest whale and largest living animal on Earth, is a baleen whale. This species may reach a length of 100 feet (30 meters) and weigh 150 tons (136 metric tons).

A structure called *baleen* is the distinguishing feature of the baleen whales. Baleen is a growth of long plates that hang in the mouth of the whale from inside the upper lip. These plates are made up of a hardened protein called *keratin*—the substance that makes up human hair and fingernails. At the back edge of each baleen plate is a fringe of course, hairlike bristles. These bristles act as a strainer, trapping small marine animals and PLANTS from water that is brought into the mouth of the whale.

Baleen whales have different feeding methods, depending on their species and the type of food they eat. A finback, humpback, or

blue whale often gulps at schools of small fish and crustaceans living in the open OCEAN. In contrast, the endangered right whale swims slowly through the ocean with its mouth open, straining plankton from the water as it moves.

For most baleen whales, the best feeding grounds are located in the extreme northern or southern oceans. However, these whales usually travel to locations near the equator to give birth to their young, called *calves*. During **calving season**, the whales may go without food for long periods of time, living only on the fat they stored while feeding in cooler waters. Much of this fat is stored in the skin in a thick layer known as *blubber*. In addition to being a source of nutrition, blubber also helps insulate the whale from the cold waters in which it lives.

 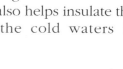

## TOOTHED WHALES

Toothed whales include the sperm whale, the orca, the narwhal, the beluga, and the smaller dolphins and porpoises, among others. In contrast to baleen whales, toothed whales have teeth that they use to capture their prey. The orca often feeds on marine mammals and BIRDS. However, most toothed whales feed mainly upon fish and squid.

To capture their prey, toothed whales need to have speed and agility. They also must have a means for finding prey in dark or cloudy waters. To assist in this activity, most toothed whales have the ability to echolocate. In addition to helping a whale locate prey, echolocation is also useful for navigation.

## WHALING AND ITS EFFECTS

Because of their large size and rich yield of meat, oil, and fat, whales have been hunted by humans for centuries. At first, whaling was done from small sailing or paddled vessels. From such vessels, whalers relied on spears, nets, or poisons to kill the animals. Using such methods, whales were killed in fairly small numbers.

In more recent years, technology has provided new methods for finding and killing whales. For much of the twentieth century, whaling has been conducted from large, fast-moving ships. Many of these ships are equipped with sonar, explosive harpoons, and factories that can process the whales as they are captured and killed. As a result of more successful hunting methods, many species of whales have not been able to reproduce at a rate fast enough to maintain their populations.

The slow-moving northern right whale was among the first whale species to near EXTINCTION. Today, only about 500 whales of this species exist in the world's oceans. Other whale species that have declined in number as a result of whaling practices include the gray whale, the fin whale, the blue whale, the humpback whale, and the bowhead whale.

In the nineteenth century, intensive HUNTING of the sperm whale began. Despite this extensive hunting, however, populations of the sperm whale are estimated to be still in the hundreds of thousands.

## CONSERVATION AND THE FUTURE OF WHALES

People and whales have a complex relationship to one another and to the marine ENVIRONMENT. In some parts of the world, people not only eat whales, but they also feed on the fish and crustaceans the whales use for food. Both humans and

**THE LANGUAGE OF THE ENVIRONMENT**

**calving season** period in which whales produce offspring.

**krill** a type of crustacean that constitutes the principal food of baleen whales.

◆ Orca, a toothed whale, feeds on such prey as fish, squid, marine mammals, and birds.

whales are PREDATORS at the top of the marine FOOD CHAIN. Thus, even if humans stop hunting whales, they will continue to compete with whales for food.

In addition to hunting whales and competing with them for food, humans create other problems for whales and their food supply. Marine organisms, including whales, are often harmed by the discharging of pollutants into the oceans. These pollutants may result from illegal, but intentional, OCEAN DUMPING; accidents that result in OIL SPILLS; or from other human activities. Thus, even as whale populations slowly recover from hunting, they may be limited in their recovery by other problems. Despite such problems, efforts to protect whales are becoming more common as people learn more about these animals.

Many people now recognize that whales are highly specialized, interesting, and intelligent animals that should be valued for their aesthetic qualities. At the same time, whales continue to be valued for their oil and protein. Human management of whales must therefore include economic, scientific, political, and ethical considerations. [See also ENDANGERED SPECIES; FISHING, COMMERCIAL; INTERNATIONAL CONVENTION FOR THE REGULATION OF WHALING (ICRW); INTERNATIONAL WHALING COMMISION; MARINE MAMMAL PROTECTION ACT; and MARINE POLLUTION.]

# Whaling

See WHALES

# Wild and Scenic Rivers Act

Legislation designed to protect rivers from being dammed or developed in other ways. In the mid-1960s, environmentalists became concerned that many free-flowing rivers of the United States were being damaged due to the construction of DAMS and other flood control projects. They feared that rivers in their natural state would soon disappear altogether. Under the Wild and Scenic Rivers Act, the construction of dams, HYDROELECTRIC POWER plants, and other projects is prohibited on and around certain designated rivers. The act also limits the extraction of MINERALS from any river in the National Wild and Scenic Rivers System.

When Congress passed the Wild and Scenic Rivers Act in 1968, it established three categories of rivers to be protected: wild, scenic, and recreational. Wild rivers are rivers that have never been developed and can be reached only by trail. Scenic rivers are mostly undeveloped but may be accessible by a few roads. Recreational rivers are developed rivers with a lot of road access.

Today, there are 152 rivers designated in the Wild and Scenic Rivers System. These rivers cover a total of 10,516 miles (16,920 kilometers). Rivers eligible for protection under this act must be undammed and have at least one other outstanding characteristic, such as a recreational, scenic, historical, or geological feature.

Individual states can request that a particular river be included in the national system. However, iden-

◆ The Wild and Scenic Rivers Act of 1968 protects some U.S. rivers from excessive damage due to damming.

tifying potential wild and scenic rivers is sometimes controversial. Predicting potential controversy, Congress made the Wild and Scenic Rivers Act flexible. For example, when a river becomes included in the national system, any preexisting LAND USE along that river, such as MINING, logging, and farming, is permitted to continue. In addition, although the act prohibits new development on designated rivers, some development is allowed, provided it is done in a way that does not harm the ENVIRONMENT. [*See also* DEPARTMENT OF THE INTERIOR; NATIONAL PARK SERVICE; and WILDERNESS ACT.]

# Wilderness

▌Undisturbed lands that are protected from the effects of human activities such as construction and agriculture. Wilderness areas are pristine and healthy ECOSYSTEMS that are relatively untouched by human use.

In response to public pressure to protect PUBLIC LANDS from excessive development, Congress passed the WILDERNESS ACT. The stated goal for this legislation, which was passed in 1964, is to "secure for the American people of present and future generations the benefits of an enduring resource of wilderness." The act defines *wilderness* as follows: "A wilderness, in contrast with those areas where man and his own works dominate the landscape, is hereby recognized as an

◆ Wilderness areas are untouched sections of land that are protected from development, such as logging and mining.

area where the earth and its community of life are untrammeled by man, where man himself is a visitor who does not remain." The Wilderness Act also makes it clear that an area is to "be protected and managed so as to preserve its natural conditions."

Wilderness areas, along with NATIONAL PARKS, NATIONAL WILDLIFE REFUGES, and NATIONAL FORESTS, are public lands. Such lands are managed by federal, state, and local governments. The federal government owns these lands; thus they can be used by the public for a variety of purposes.

Wilderness areas are the most protected of the public lands. Unlike FORESTS and wildlife refuges that allow some limited amounts of development such as MINING, logging, GRAZING, farming, and OIL DRILLING, wilderness areas can be used only sparingly. Activities such as hiking, camping, fishing, and boating are allowed in wilderness areas. However, motorized equipment of any kind is prohibited. The Wilderness Act also prohibits the building of roads and other structures, as well as commercial activities, such as harvesting timber and large-scale mining.

## NATIONAL WILDERNESS PRESERVATION SYSTEM

The 1964 Wilderness Act established the National Wilderness Preservation System (NWPS), the system of wilderness lands in the United States. Several federal agencies, including the FISH AND WILDLIFE SERVICE, the U.S. FOREST SERVICE, the NATIONAL PARK SERVICE, and the BUREAU OF LAND MANAGEMENT manage wilderness lands for public use. Wilderness areas can be found throughout the United States. When the NWPS began, only 54 wilderness areas were recognized. Today, there are over 500 wilderness areas covering almost 95 million acres (38.5 hectares). All of these areas have been identified and are protected. Many states have also passed laws designating their own forms of wilderness areas.

## WILDERNESS AREAS AND HUMAN SOCIETY

In addition to providing areas for enjoyment and recreation for humans, wilderness areas also provide HABITAT for countless numbers of wild animals and PLANTS. Before passage of the Wilderness Act in 1964, people debated about whether or not to preserve these ecosystems. Most people would agree that some land in the United States should be preserved and maintained in a relatively undisturbed state. Today, however, the debate is more about how much land to protect. Opponents argue that "locking up" too much land as wilderness is not fair to business and hurts the economy. They argue that as the population of the United States grows, more land will be needed to build roads, bridges, and homes, and more NATURAL RESOURCES, such as MINERALS and timber, will be needed to complete these types of projects and to meet other needs of people.

Supporters of the Wilderness Act, on the other hand, argue that many of the animals and plants in designated wilderness areas are near EXTINCTION. They feel that ecosystems should be preserved to ensure the survival of these SPECIES. Supporters also point out that even though wilderness areas are protected from the damaging effects of construction and motor vehicles, they are still threatened by ACID RAIN, fire, WATER POLLUTION, and other harmful conditions resulting from human activities. Finally, supporters agree that compared to the overall amount of public lands, relatively few areas are classified as wilderness. They argue that these remaining lands should be preserved for human enjoyment before they are gone forever. [*See also* BIODIVERSITY; ENDANGERED SPECIES; ENVIRONMENTAL ETHICS; FOREST SERVICE; FRONTIER ETHIC; HABITAT LOSS; LAND USE; MULTIPLE USE; OVERGRAZING; SUSTAINABLE DEVELOPMENT; WILD AND SCENIC RIVERS ACT; WILDERNESS SOCIETY; WILDLIFE CONSERVATION; and WILDLIFE MANAGEMENT.]

# Wilderness Act

▮The 1964 Wilderness Act signed into law by President Lyndon B. Johnson that created the National Wilderness Preservation System, a system of PUBLIC LANDS that now contains more than 100 million acres (40 million hectares) of protected WILDERNESS around the United States. Wilderness areas are defined as lands that are undisturbed and protected from human activities. According to the act, wilderness is

◆ The goal of the Wilderness Act is to protect the country's wilderness areas for human enjoyment.

land that exists in its natural, undisturbed condition.

The purpose of the Wilderness Act is to protect the few remaining untouched regions of the United States from damage due to human activities. The law prohibits construction of roads and buildings, as well as industry such as farming and MINING, on lands designated as wilderness. Low-impact activities such as hiking, camping, swimming, and boating are allowed as long as no motorized equipment is used.

The Wilderness Act arose after many years of public pressure on Congress to protect some of the natural, undisturbed public lands in the United States for human enjoyment.

Its roots lie in the pioneering efforts of conservationist Aldo LEOPOLD, who, in 1921, as an employee of the U.S. FOREST SERVICE, proposed that certain parts of the NATIONAL FORESTS be preserved as wilderness areas. When the Wilderness Act was passed in 1964, about 50 areas totaling nearly 9 million acres (3.7 million hectares) of untouched land were immediately designated as wilderness. Today, over 500 separate wilderness areas are protected under this law.

A recent addition to the National Preservation System came in 1994, when President Clinton signed the California Desert Protection Act, which set aside a large area of untouched DESERT in California as wilderness.

The majority of wilderness land in the United States is in the state of Alaska. However, wilderness areas are found in all 50 states. Wilderness lands include NATIONAL FORESTS managed by the U.S. Forest Service, NATIONAL PARKS managed by the NATIONAL PARK SERVICE, NATIONAL WILDLIFE REFUGES managed primarily by the U.S. FISH AND WILDLIFE SERVICE, and other areas under the care of the BUREAU OF LAND MANAGEMENT. [*See also* ENVIRONMENTAL ETHICS; FRONTIER ETHIC; LAND USE; MULTIPLE USE; SUSTAINABLE DEVELOPMENT; WILDERNESS SOCIETY; WILDLIFE CONSERVATION; and WILDLIFE MANAGEMENT.]

# Wilderness Society

**❙A**n organization based in Washington, DC, that is concerned with preserving, or keeping, the ENVIRONMENT in its pristine condition. The Wilderness Society was founded in 1935 by wildlife manager and naturalist Aldo LEOPOLD and Robert Marshall, an officer in the U.S. FOREST SERVICE. The society was formed as a result of a dispute about how PUBLIC LANDS of the United States should be treated. The dispute over this issue continues to this day.

Preservationists, including members of the Wilderness Society, believed that large areas of public land should remain as unspoiled HABITATS, protected from MINING, tree-cutting, GRAZING, and most other commercial uses. The preservationists believed such areas should be enjoyed by people and passed on unspoiled for research and enjoyment by future generations. In contrast, others saw public lands as a national resource that should be used for economic growth. They believed the land

should be mined for MINERALS, the trees should be harvested, and the grass grazed.

The government is viewed as merely a particularly farsighted manager, harvesting trees for SUSTAINED YIELD instead of cutting them all down at once and letting the SOIL erode as private owners sometimes do. Many conservationists believe in "multiple use and sustained yield," a theory that still guides the management of the national FORESTS of the United States. [*See also* BIODIVERSITY; CONSERVATION; WILDERNESS; and WILDERNESS ACT.]

# Wildlife

**❙P**LANTS and animals that live on the land and in the watery areas of Earth. Wildlife is the working network of organic existence on our planet.

Wildlife is important to human beings. By studying it, scientists have learned much about what people need to live and how human illnesses can be treated. Many wild plants yield new medicines. In addition, GENES from wild plants can be used to improve existing crop plants by making them more disease resistant or more productive. Scientists now know how to transplant genes from one SPECIES to another; thus, they have the ability of changing the characteristics of species. Most importantly, wildlife is a significant part of our ECOSYSTEM, without which many human activities, and perhaps, human life, are not possible.

## THE IMPORTANCE OF HABITAT

Plants and animals find what they need to survive in their HABITATS. An organism's habitat provides these:

**1.** Space for normal growth, movement, or territorial behavior. For example, some animals migrate with the seasons to mate or search for food.

**2.** Food and water. For example, BIRDS that feed on seeds and fruit need seed- and fruit-producing plants in their habitat. Plants need MINERALS in the SOIL.

**3.** Places for breeding and raising young. Some FISH require the rocky bed of a stream for laying their eggs. The osprey, or fish hawk, needs a tall tree with large branches at the top where it can build its nest and safely raise its hatchlings.

**4.** Cover or shelter. The wide branches of pine and fir trees in the FOREST shelter small MAMMALS and birds. Bears and other animals **hibernate** during winter in caves and crevices. Snakes, which cannot produce heat naturally, warm themselves on and under rocks that absorb heat.

Within a habitat, plants and animals live together and interact. Organic waste products decompose and enrich the soil in which plants grow. Earthworms tunnel through the soil, creating spaces through which water and OXYGEN can circulate. Birds and flying INSECTS feed on plants and spread plant **pollen**, thus helping the plants to reproduce. Fish eat ALGAE, and seabirds eat the fish, creating a simple FOOD CHAIN through which energy moves through an ECOSYSTEM.

## HUMAN THREATS TO WILDLIFE

Early human beings lived closely with wildlife. Animal flesh provided food. Animal skins were used for clothing and shelter. Animal bones were used to make tools and weapons. Some animals—such as dogs, sheep, horses, and cattle—were **domesticated**.

Today, people shelter themselves in buildings and produce food on farms and ranches, safe from other animals. Many human activities, however, still result in damage to wildlife and wildlife habitats.

## DWINDLING HABITATS

Any disturbance of a habitat can prove fatal to the wildlife that depends upon it. Destruction of habitat occurs as people build highways and cities, convert RAIN FORESTS to ranch lands, and pollute water. Damage to habitats, which has already caused the EXTINCTION of many species of plants and animals, is probably the greatest threat to wildlife today.

Over the past 125 years, about 57 species of birds, mammals, and fish have become extinct in the United States. Worldwide, three or more animal or plant species become extinct every hour, according to natural scientist Edward Osborne WILSON. The toll on plant species is believed to be the same or higher. The extinction rate for invertebrate species is thought to be much higher. At this rate, by the year 2000, 10% of all Earth's species will be gone.

In recent years, people have begun to become aware of their

◆ The forest habitat provides the white-tailed deer with its essential needs, such as food and water, breeding and living space, and shelter.

obligation to preserve wildlife. Nature reserves and NATIONAL PARKS have been designated as NATIONAL WILDLIFE REFUGES. Laws protect some animals and plants, but it may be too late for many species. Although scientists are working to save them, many species such as the black-footed ferret, the GORILLA, and the blue WHALE, are all on the brink of extinction. [*See also* BIODIVERSITY; CLEAR-CUTTING; CONSERVATION; CONVENTION ON INTERNATIONAL TRADE IN ENDANGERED SPECIES OF WILD FLORA AND FAUNA (CITES); ENDANGERED SPECIES; ENDANGERED SPECIES ACT; MASS EXTINCTION; WILDLIFE CONSERVATION; WILDLIFE MANAGEMENT; and WILDLIFE REHABILITATION.]

# Wildlife Conservation

▶ The **sustainable** management of Earth's PLANTS and animals. CONSERVATION means not only protecting the lives of plants and animals but also protecting their HABITATS. A habitat is the natural ENVIRONMENT of an organism and contains everything an organism needs to live.

WILDLIFE is vital to human existence. Plants and animals produce many things we need to live—food, shelter, and air. The study of plants and animals has also produced much vital information about the human body and human illnesses. In addition, plants yield many medicines. In fact, one-sixth of all medicines used for treatment of stomach ulcers, heart disease, and infections are produced by using tropical plants. Nearly 1,500 tropical plants are believed to contain anti-CANCER compounds.

## OVEREXPLOITATION

Wild plants and animals were once considered to exist for human use. In many cases, human activity has caused the EXTINCTION of a SPECIES. Before the nineteenth century, the PASSENGER PIGEON was the most common bird in the United States. Flocks of the large BIRDS blackened the skies. However, passenger pigeons were also used as food and hunted for sport. Hunters shot these birds by the thousands. In the early 1900s, overhunting and HABITAT LOSS led to the extinction of the passenger pigeon.

Many plant species have also become **extinct**. Among them are several species of cacti. Cacti, which are often harvested for plantings, seldom thrive in cultivated situations. Many cactus species no longer exist or exist in only small numbers because of the activities of collectors.

Animals are threatened by collectors, too. In the mid-1800s, many fashions made use of feathers as decorations. Hundreds of thousands of egrets, heron, swans, gulls, swallows, birds of paradise, albatrosses, and ostriches were killed for their feathers. A law passed in 1913 ended the feather trade. However, trumpeter swans and herring gulls were nearly wiped out before this law was passed. Today, many other animal species are threatened by the activities of people. Tropical birds, chimpanzees, and monkeys are captured in their RAIN FOREST habitats and shipped to the United States and Europe, where they are sold to pet stores and animal collectors. Rhinoceroses are hunted for their horns and ELEPHANTS for their valuable tusks. As a result, many of these animals are threatened with extinction.

## HABITAT DESTRUCTION

The greatest danger to wildlife is the destruction of habitat. The clearing of land for homes, farmland, and highways destroys habitat. When a WETLAND is filled in or a GRASSLAND plowed, many kinds of wildlife suddenly have no shelter, no food, no water, and no place to raise their young.

The human need for lumber may result in the destruction of a FOREST. In the Pacific Northwest, almost all of the OLD-GROWTH FORESTS have been harvested. These forests are home, pantry, and nursery for the NORTHERN SPOTTED OWL, pileated woodpecker, pine marten, elk, and Pacific yew. The bark of the Pacific yew contains an important anti-cancer compound.

Some Pacific Northwest forest areas have been clear-cut. This

---

**THE LANGUAGE OF THE ENVIRONMENT**

**extinct** no longer existing.

**sustainable** relating to a method of using a resource without depleting it.

practice often leaves behind barren ground that is at risk of EROSION. Rain washes the SOIL into streams, where it collects and destroys many kinds of organisms, including SALMON. Salmon migrate from the sea to the forest streams to lay their eggs on rocky stream beds. When rainwater carries soil particles into streams, the salmon eggs and small hatchlings are often smothered by the SEDIMENT.

Habitats can also be damaged by POLLUTION. For example, DIOXINS are currently found in small amounts in almost all the animals in the Gulf of Mexico. Through BIOACCUMULATION, however, top PREDATORS like the BALD EAGLE are building up high levels of dioxins. Some researchers think that these dioxins are responsible for some recent nesting failures by bald eagles in that area.

The introduction of an alien, or exotic, species can also threaten the survival of NATIVE SPECIES. Removed from its natural environment, an EXOTIC SPECIES may find few enemies and other BIOLOGICAL CONTROLS on its POPULATION GROWTH in its new environment. For example, pigs and goats were brought to the Hawaiian Islands by explorers. The animals, intended to provide meat for ships' crews, were allowed to run wild. Today, these exotic species are destroying many of the islands' plants.

## EXTINCT, ENDANGERED, THREATENED

Human activity has directly caused the EXTINCTION of many species. Between A.D. 1 and A.D. 1800, one MAMMAL species became extinct every 55 years. During that time, a

◆ Wild hogs are strong, fierce animals that live in forests and jungles in many parts of the world, including Europe, Asia, and Africa.

total of 35 mammal species were lost forever. Human activities have increased the damage to habitat. They have also increased the rate of extinction. Some scientists estimate that 57 species of mammal, bird, and fish have died out during the past 125 years. Many believe the extinction rate for VERTEBRATES (animals with a backbone) and plants is about one or two per year. The extinction rate for INVERTEBRATES is thought to be much higher.

In the United States, as of February 1994, 759 species had been legally listed as endangered and 196 as threatened. In addition, over 4,000 species are under consideration for being placed on either

the threatened or endangered species list.

## CONSERVATION MEASURES

In the late nineteenth century, people became aware of and concerned about the plight of wildlife. In 1872, the U.S. Congress set aside YELLOWSTONE NATIONAL PARK as the first NATIONAL WILDLIFE REFUGE. In 1916, the U.S. NATIONAL PARK SERVICE was formed. Its purpose was to watch over the wildlife in lands set aside as NATIONAL PARKS.

Conservation measures include laws that protect the lives and habitats of wildlife. HUNTING and fishing laws limit the number, size, or age

◆ The trumpeter swan is an endangered species that is recovering due largely to conservation efforts.

of animals that can be hunted. Animals may not be hunted during those times of the year when they are mating and raising offspring.

The ENDANGERED SPECIES ACT of 1973 made it illegal to kill or capture threatened or endangered species. This law also protects habitat. No project such as a highway or housing development can interfere with the habitat of a threatened or endangered species. The law was so successful at protecting alligators, which had nearly been hunted out of existence in Florida, that limited hunting is now permitted to keep their numbers in check. The law was recently enforced to protect the northern spotted owl by restricting the number of trees that could be cut in the forests of Washington State.

International laws protect life in the sea. In the Pacific, California gray WHALES, which had been hunted until only a few thousand remained, were saved by hunting bans enforced by the INTERNATIONAL WHALING COMMISSION. Now number-ing 24,000, the whales have been removed from the endangered species list.

Laws banning pollution also protect habitat. A 1972 law banning the use of the pesticide DDT is credited with saving the bald eagle from extinction. Since the law was passed, the bald eagle population has increased from 400 to 4,000 in the 48 continental states. Recently, it has been proposed that this bird be removed from the endangered species list and placed on the threatened species list. Population increases have also been noted in other birds once threatened by DDT. These species include brown pelicans, osprey, and peregrine falcons.

In some cases, the population of an endangered animal species is so low that scientists decide to remove it from its habitat for its own protection. In captivity, the animal is allowed to mate and produce young. Small numbers of the species are then reintroduced into protected environments. This is a new idea, and thus far results are uncertain. The last 17 red wolves were captured for breeding. Scientists later released 60 of the animals into habitats in the southeastern United States. The news is not as good for the 5-foot-tall (1.5 meters) whooping crane, which had been ruthlessly hunted and pushed out of its wetland habitat by construction projects. The birds do not breed well in captivity, and scientists have abandoned efforts to produce flocks. However, habitat protection seems to be helping the whooping crane population.

Of 20,000 plant species in the United States, 4,000 are endangered or threatened. This large number seems to indicate that entire plant communities are being destroyed by construction or pollution. Computers are now being used to map states' plant and animal communities area by area. Areas inhabited by wide varieties of organisms can be designated as wildlife preserves. This method makes it possible to save many different species at the same time.

Since the passage of the Endangered Species Act, 21 species have been removed from the endangered species list. Some of these were removed because they were extinct. Others, like the American alligator and the brown pelican, were taken off the list because their populations have managed to recover. But with over 4,000 species waiting to be considered for protection in the United States alone, some scientists have begun to think that not all endangered species can be saved from extinction. Choices may have to be made about which species should be saved. Whatever the decisions, the Earth will lose species it can never regain. [*See also* ANIMAL RIGHTS; ARCTIC NATIONAL WILDLIFE REFUGE (ANWR); BIODIVERSITY; CAPTIVE PROPAGATION; CARRYING CAPACITY; CLEAR-CUTTING; CLIMAX COMMUNITY; CONVENTION ON INTERNATIONAL TRADE IN ENDANGERED SPECIES OF WILD FLORA AND FAUNA (CITES); DEBT FOR NATURE SWAPS; DEEP ECOLOGY; DEFORESTATION; ENDANGERED SPECIES; ENVIRONMENTAL PROTECTION AGENCY (EPA); EVERGLADES NATIONAL PARK; FISH AND WILDLIFE SERVICE; GALÁPAGOS ISLANDS; HABITAT LOSS; INTERNATIONAL UNION FOR THE CONSERVATION OF NATURE (IUCN; NATIVE SPECIES; NORTHERN SPOTTED OWL; PET TRADE; RESTORATION BIOLOGY; WETLANDS PROTECTION ACT; WILDERNESS ACT; WILDERNESS SOCIETY; and WILDLIFE MANAGEMENT.]

◆ Snow geese are found in great numbers at the National Wildlife Refuge in New Mexico.

# Wildlife Management

❚ Practices that promote the health, survival, and sustainable use of wild animal populations. Animal populations are managed for several reasons. Some SPECIES are managed because they may be rare and in danger of EXTINCTION. Others are managed to promote and protect the health of ECOSYSTEMS. Still others are managed for the benefit of humans. Wildlife management practices fall into three main categories:

**1.** development and enforcement of laws to protect wild animals from overhunting and HABITAT destruction;

**2.** improvement and restoration of habitats to maintain the health and the size of WILDLIFE populations; and

**3.** management and CAPTIVE PROPAGATION of rare and ENDANGERED SPECIES to increase population sizes.

## PROTECTING NATIVE ANIMAL SPECIES

Most countries recognize the importance of BIODIVERSITY. Many have laws designed to protect and promote the well-being of NATIVE SPECIES. In the United States, wildlife protection laws have existed since the turn of the century.

### Hunting Regulations

Early wildlife laws were primarily developed in response to the overhunting of BIRDS, such as PASSENGER PIGEONS, ducks, and geese. Some birds were hunted simply for recreation and enjoyment. However, most were taken in large numbers for sale on the open market.

In 1900, Congress passed the Lacey Act, which made it illegal to ship wild animals across state lines. Several years later, in 1918, the MIGRATORY BIRD TREATY was passed, making it illegal to hunt migratory birds along major flyways. Other protective laws have since outlawed commercial HUNTING.

The regulation of recreational hunting is also a protective measure for managing wild animal populations. Hunting is permitted in limited amounts only during certain times of the year. In addition, all states have laws that require hunters to purchase hunting permits, which help pay for the salaries and equipment used by game wardens.

The U.S. FISH AND WILDLIFE SERVICE (FWS) has the responsibility of making sure states enforce hunting laws. In 1973, with the passage of the ENDANGERED SPECIES ACT, the FWS was also charged with the task of protecting threatened and ENDANGERED SPECIES. These are species that are likely to become extinct unless they are protected. The Endangered Species Act prohibits hunting and any other activity that may jeopardize an endangered or threatened species.

## National Wildlife Refuges and Sanctuaries

The Fish and Wildlife Service also manages a system of NATIONAL WILDLIFE REFUGES that offer protected habitats for animal species. There are over 400 refuges, covering an area of 90 million acres (36 million hectares) in the National Wildlife Refuge System. Most have been established to protect the breeding and wintering grounds of migrating birds. Others have been developed to protect the habitats of endangered and threatened species.

### HABITAT MANAGEMENT

The enforcement of laws protecting animal populations is an essential part of wildlife management. However, in most cases this isn't enough. The habitats of wildlife populations must also be maintained so populations remain healthy. Protecting habitats from destructive human activities, such as MINING, OIL DRILLING, GRAZING of animals, and logging, is perhaps the most important aspect of habitat management. Establishing pro-

tected areas, such as WILDERNESS areas, NATIONAL PARKS, and wildlife refuges, has been a successful strategy for preserving animal habitats.

Improving and restoring habitats is another aspect of habitat management. Most national wildlife refuges and state-managed wildlife sanctuaries practice habitat improvement. These techniques may involve building structures to provide species with nesting sites and shelter from PREDATORS. They may also include increasing water supplies by artificial means. Habitat restoration refers to methods that attempt to restore damaged habitats to their original, predisturbed conditions.

### Regulation of Population Size

Habitat management techniques are used to maintain and increase the size of animal populations. This is particularly true for game animals, such as ducks, geese, and deer. Using these methods, wildlife managers can regulate population sizes for the benefit of recreational hunters. Game production is generally increased by keeping populations just below the CARRYING CAPACITY of the ENVIRONMENT. For instance, regulating the size of a deer population may involve killing old, sick animals, increasing water availability, or cutting mature trees to allow the growth of young plants, a source of food for deer.

### CAPTIVE PROPAGATION

For rare and endangered species, protection from hunting and habitat destruction is often not enough to save a species from extinction.

Many animals, such as CALIFORNIA CONDORS and red wolves, have been saved from extinction through captive breeding programs. Many ZOOS and wildlife parks establish such programs to increase population sizes. Captive breeding programs are expensive. However, using new technologies such as artificial insemination and embryo transfers, these programs can be quite successful. Once a captive population has increased sufficiently in size, animals can be introduced gradually back into the wild. [See also ARCTIC NATIONAL WILDLIFE REFUGE (ANWR); CONSERVATION; CONVENTION ON INTERNATIONAL TRADE IN ENDANGERED SPECIES OF WILD FAUNA AND FLORA (CITES); EVERGLADES NATIONAL PARK; FOREST; HABITAT LOSS; INTERNATIONAL UNION FOR THE CONSERVATION OF NATURE (IUCN); MARINE MAMMAL PROTECTION ACT; WILDERNESS ACT; WILDLIFE CONSERVATION; and WILDLIFE REHABILITATION.]

# Wildlife Rehabilitation

Care and treatment of injured or orphaned WILDLIFE. People often find injured wildlife. Animals are injured by cats and dogs, AUTOMOBILES, poisons, and through natural predation and illness. BIRDS commonly hit windows, telephone lines, and support wires on towers. Birds migrating at night often fly into buildings and radio towers. People also find young wild animals that either have been orphaned or appear to be orphaned.

Wildlife rehabilitators try to treat, care for, or release injured and orphaned wildlife. Some rehabilitation centers operate as part of ZOOS, museums, or nature centers, while others are run by individuals as a hobby or nonprofit business. Being a wildlife rehabilitator requires extensive knowledge of wildlife and training by a veterinarian or other rehabilitator. It also requires an enormous investment of time, especially in the raising of young birds. Some species require feeding every 20 minutes, all day long. Permits from state fish and wildlife organizations are also needed in order to legally hold and treat wildlife. Special federal permits are needed in order to treat birds of prey.

Many wildlife experts have doubts about the effectiveness of most wildlife rehabilitation programs. If orphaned animals grow up associating people with food or are cared for by people from a young age, they may seek people out when looking for food. Unless the rehabilitator is very careful, an animal can also end up not having all the skills it needs to survive in the wild and may thus have trouble finding food or evading PREDATORS. There is very little information available on the survival rates of injured animals that are treated by rehabilitators and then released. Wildlife rehabilitators, however, tend to believe very strongly in the value of their work and are often motivated by a concern for the suffering of injured animals. [*See also* WILDLIFE CONSERVATION and WILDLIFE MANAGEMENT.]

# Wilson, Edward Osborne (1929– )

▶ **A**merican biologist at Harvard University who is an expert on social INSECTS. Wilson is responsible for significant research in biogeography—the study of the distribution of SPECIES, and in sociobiology—the study of the genetic basis of social behavior in animals, including humans. A two-time winner of the Pulitzer Prize for nonfiction, Wilson's works include *The Ants* (1990), the most complete book ever written on these insects, and *Naturalist* (1994), an autobiography.

Wilson's earlier works include *Sociobiology: The New Synthesis* (1975), which started an intense controversy in the scientific community. To clarify the relationship between behavior in humans and similar behavior in other animals, Wilson suggested that certain aspects of human behavior, such as aggression, kindness, or dividing work based on sexual differences, were determined by genetic makeup.

In the 1980s, Wilson focused attention on the threatened EXTINCTION of many of Earth's species. His book, *The Diversity of Life* (1992), deals with the vast BIODIVERSITY of organic and microorganic species in our world. [*See also* GENETIC DIVERSITY; SPECIES DIVERSITY; and SUCCESSION.]

# Wind Power

▶ **T**he energy in wind, which can be harnessed by windmills. Windmills can be used to do work directly, such as powering a water pump. However, a more common use of windmills today is to generate ELECTRICITY.

Windmill generators have a relatively long history, especially in areas with no other source of electricity. In the 1930s, for instance, Zenith Radio sold thousands of "Windchargers" to people living in remote parts of the United States. These windmills powered AUTOMOBILE generators that were used to recharge batteries. The batteries ran radios, the only contact with the outside world for isolated farms without electricity or telephones.

The technology of windmill generators is now well developed, and the cost of the electricity they produce is lower than that from many other energy sources. There

◆ Wind turbines at Tehachapi, California, are part of an experimental project to use wind power.

is enough energy in the windiest spots on Earth to generate more than ten times the electricity now used worldwide. However, wind-generated energy fills only a tiny portion of the world's energy needs.

Why is wind power not more popular? The main reason is that windmill generators must be located where there is enough wind to make them economical. They also have other disadvantages. They are large, often unattractive additions to the landscape. BIRDS often fly into them, and the blades of large windmills may interfere with microwave communications.

Despite their disadvantages, windmill generators are useful in windy areas, such as some of the Caribbean islands and parts of the United States. In these areas, they provide electricity at a reasonable cost. [*See also* ALTERNATIVE ENERGY SOURCES.]

# World Bank

Also called the International Bank for Reconstruction and Development. The World Bank was founded in 1946 by the United Nations (U.N.). The World Bank finances the economic development of countries belonging to the United Nations.

Funds for the World Bank come from U.N. members. These funds are lent to private utilities for building electric power plants and systems, as well as for transportation and water supply lines. Other projects financed by the World Bank include agricultural and rural development, inland waterways, airlines, airports, telecommunications systems, pipelines, highways, roads, IRRIGATION and flood control, land clearance and improvement, crop processing and storage, LIVESTOCK development, and FORESTRY.

The World Bank has been criticized because some of its projects have damaged the ENVIRONMENT. Recently, the organization has put more emphasis on the environmental effects of projects it finances. [*See also* CONSERVATION; DEBT FOR NATURE SWAPS; GREEN POLITICS; and GREEN REVOLUTION.]

# X

## Xenobiotic

▶ Any organic or inorganic substance that is foreign, and usually harmful, to organisms or biological systems. Derived from the Greek words *xeno* ("foreign") and *bios* ("life"), xenobiotics affect individuals, communities, and ECOSYSTEMS.

centrations, they are not xenobiotics. Because life-essential trace amounts of these substances are usually found in the body, they are not considered "strangers" to living things. [*See also* COMMENSALISM; HAZARDOUS WASTE; HEAVY METALS POISONING; MUTUALISM; PARASITISM; SYMBIOSIS; and TOXIC WASTE.]

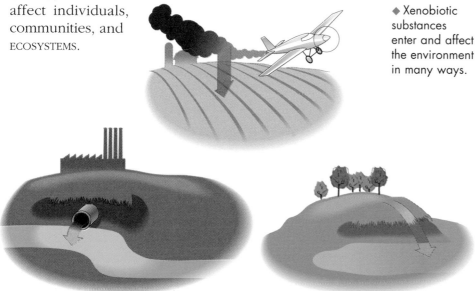

◆ Xenobiotic substances enter and affect the environment in many ways.

Nutrient RUNOFF and BIOACCUMULATION have xenobiotic effects at the ecosystem level. Parasites may be considered xenobiotic at the community level. Foods, drugs, poisons, and chemicals such as MERCURY are xenobiotic at the individual level. Depending on their concentrations, these substances may be toxic or cause CANCER. Although some substances, such as chromium and zinc, may be toxic or cause cancer in large con-

## X Rays

▶ Intense energy or RADIATION produced by atomic particles called *electrons*. Quantities of x rays can be released from electrons by blasting a stream of electrons onto a metal target within a glass tube, enabling medical experts and others to put them to use. X rays travel through matter at different rates and are used to produce interior images of living bodies, as well as of machinery and metal parts.

X rays can damage living tissue, and many safeguards are used by health professionals to protect against this. The tissue-damaging quality of this radiation is used to kill cancer cells. The x rays are targeted at the location of the CANCER in the person's body. Other parts of the patient's body are shielded with LEAD, which will not admit x rays.

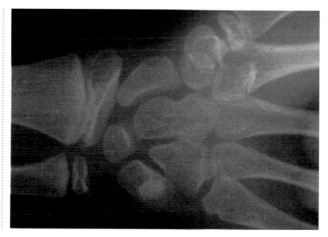

◆ X rays are used to see inside the body, providing doctors with information about patients' illnesses.

# Y

## Yellowstone National Park

The oldest and largest NATIONAL PARK in the lower United States (excluding Alaska). Established by Congress in 1872, Yellowstone National Park contains 2,200,000 acres (880,000 hectares). It is located in the northwestern corner of Wyoming and extends into Idaho and Montana. As large as it is, however, Yellowstone is not large enough to support viable long-term populations of many SPECIES, including the threatened GRIZZLY BEAR.

◆ Mammoth Hot Springs is found in Yellowstone National Park, the oldest and largest national park in the lower United States.

◆ The "Old Faithful" geyser brings many tourists to Yellowstone National Park.

The park is part of a greater Yellowstone ECOSYSTEM encompassing from 14 to 19 million acres (7 to 8 million hectares), but the concept of ecosystem management did not exist in the 1800s. Thus, the park's boundary cuts straight across mountain ranges, WATERSHED, and wildlife HABITAT. The park lacks what has been called a fundamental feature of a model bioreserve: buffer zones. Areas adjoining the park, such as the Targhee National Forest in Idaho, have been severely damaged by heavy CLEAR-CUTTING.

The sheer volume of human visitors to the park has created problems within the park itself. Automobile traffic, first allowed in 1915, now strains the park's road system. Estimates of the cost of repairing the roads have ranged into the hundreds of millions of dollars. Other sources of controversy involving Yellowstone in recent times include fires that destroyed 1 million acres (about 400,000 hectares) in 1988 and the reintroduction of the timber wolf to the area.

John Muir lived in the Yosemite Valley during the 1880s. He spent a great deal of time studying the PLANTS, animals, and geology of the region. He wrote about his studies and became well known as a writer. He conceived the idea of creating a Yosemite National Park. At his urging and that of other naturalists, the U.S. Congress passed a law in 1890 creating Yosemite National Park, along with Sequoia and General Grant Parks.

Today, the preserve has more than 220 SPECIES of BIRDS and more than 60 species of MAMMALS in the FORESTS and mountains of Yosemite, including deer, squirrels, chipmunks, and black bears. The park is also HABITAT for more than 1,300 varieties of plants and 30 kinds of trees, including groves of giant sequoias (*Sequoiadendron giganteum;* big trees), some of which are thousands of years old. The Mariposa Grove, for example, contains the Grizzly Giant Tree, whose trunk measures more than 34 feet (10 meters) in diameter at the bottom. [*See also* NATIONAL PARK SERVICE.]

# Yosemite National Park

▌A wilderness preserve located in the scenic mountain area in the Sierra Nevada Range of central California. The park contains several stands of ancient sequoias. It is also a sanctuary for black bear and deer.

Yosemite Valley has a number of natural attractions, including sheer rock walls, huge domes and peaks, more than 300 lakes, and the world's finest collection of waterfalls. The park covers 761,320 acres (308,106 hectares) of WILDERNESS, including Yosemite Valley, which is approximately 4,000 feet (1,200 meters) above sea level. Yosemite National Park became one of the earliest NATIONAL PARKS to be created by the U.S. Congress. Much of the work to preserve the Yosemite Valley was done by John MUIR, a naturalist, conservationist, and writer.

# Z

## Zebra Mussel

*See* EXOTIC SPECIES

## Zero Population Growth

▶ A condition in which the number of people in a given area remains stabilized at its current level of growth over several generations. Birth rate and **immigration** rate increase population size. The death rate and **emigration** rate decrease population size. Zero population growth is achieved when, over time, the population increase from births and immigration is equal to the decrease from deaths and emigration.

Worldwide, the human population has the potential to reach twice its present size in the next 40 years if the current growth rate continues. Many people believe that most of today's environmental problems are a direct result of this growth, making zero population growth a desirable goal. A few countries, such as Italy, Germany, and Japan, have attained zero population growth levels, but many others, such as China and the United States, have not. [*See also* AGE STRUCTURE; CARRYING CAPACITY; EXPONENTIAL GROWTH; FERTILITY RATE; INFANT MORTALITY; and POPULATION GROWTH.]

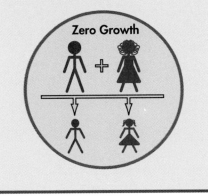

THE LANGUAGE OF THE ENVIRONMENT

**emigration** the act of moving in which people leave an area for another place.

**immigration** the act of moving in which new people move into an area.

Zero Growth

## Zone of Saturation

▶ The region of Earth's crust where all the pore spaces in SOIL and rock are completely filled with water. The water that occupies the zone of saturation is called *groundwater*.

The thickness of the zone of saturation and the amount of groundwater stored there varies, depending on local geography and the nature of the underground material. Material containing a high percentage of pore spaces per unit volume can hold a great amount of groundwater, if the pore spaces are large enough to allow the water to pass through easily.

Water in the zone of saturation can become contaminated by buried solid and TOXIC WASTES, strong acids, and agricultural RUN-OFF. In coastal areas, overpumping of groundwater may deplete the water supply and result in SALTWATER INTRUSION. [*See also* AQUIFER; CLEAN WATER ACT; HYDROLOGY; INFILTRATION; LANDFILL; LEACHING; SAFE DRINKING WATER ACT; SEWAGE; WATER POLLUTION; WATER QUALITY STANDARDS; and WATER TABLE.]

◆ Wells take up groundwater from the zone of saturation. As more and more people use this water supply, the water table drops.

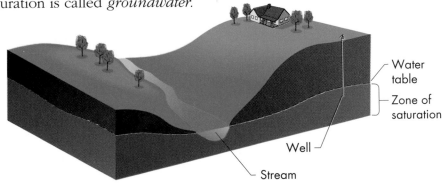

Water table
Zone of saturation
Well
Stream

◆ Zoos in which animals can live in artificial habitats similar to their real homes allow the animals more freedom and visitors an opportunity to see them in a more natural setting.

# Zoo

▌Common term used for a park, garden, or other site where wild animals, and possibly domestic animals, are housed in captivity. The word *zoo* was first used in the late nineteenth century to refer to London's zoological gardens.

◆ Cheetah cubs are reared by their mothers.

## ZOOS: PURPOSE AND FORM

A zoo is a place where animals can be observed and studied. Most zoos provide educational programs for the schoolchildren and the adults in their area. Most zoos contain a variety of animals. However, a few zoos specialize in the display of animals of one group, such as only primates or only tropical BIRDS. Some zoos include marine MAMMALS in aquariums, along with other marine life exhibits. Animals usually receive more care in zoos than in nature reserves or sanctuaries because zookeepers are better able to check on each animal's well-being.

Zoos come in many forms. Some are set up as buildings in which animals are exhibited in indoor cages. Some have buildings with indoor cages and outdoor enclosures. The use of outdoor open-range, cageless exhibits with moats or fences to keep people from the animals was first used by the Zoological Society of London in 1932. In other zoos, called *safari parks,* visitors travel by car through regions where wild animals live in somewhat natural ENVIRONMENTS.

## ZOO HISTORY

No one knows exactly when the first zoo was set up. It was possibly around the time humans first began the domestication of animals. Records show that as early as 4500 B.C., pigeons were kept in captivity in Mesopotamia. By 2500 B.C., ELEPHANTS were partially domesticated in India. Wen Wang, the ruler of China around 1000 B.C., set up a 1,500-acre (607-hectare) zoo called Ling-Yu, or Garden of Intelligence. Centuries later, during his A.D. 1519 explorations, Cortés discovered an extraordinary zoo in Mexico. The zoo was so immense that 300 keepers were required to care for all the mammals, birds of prey, and REPTILES. In the United States, the National Zoological Park in Washington, DC, was founded by Congress between 1889 and 1890 and is maintained by public funds.

## BREEDING ZOOS

After World War II, many zoos became breeding centers for endangered animal SPECIES. For example, when it was discovered in 1947 that only 50 nene, or Hawaiian geese, still existed, some of the birds were captured and sent to a zoo in England. There the nene thrived, and in 1951 a gander was hatched. The birds continued to breed successfully, and those that hatched were sent to other European zoos to lower the chances of losing all of the endangered birds from disease if the nene were housed in only one place. Now, CALIFORNIA CONDORS and other threatened or ENDANGERED SPECIES are bred in zoos around the world. Many scientists hope one day to restock the wild with many species once on the verge of EXTINCTION. [*See also* BIOME; GENETIC ENGINEERING; GIANT PANDA; and KEYSTONE SPECIES.]

# Zooplankton

◗ Very small animals that live at or near the surface of lakes or OCEANS. Unlike the tiny marine algae called PHYTOPLANKTON, zooplankton do not make their own food by PHOTOSYNTHESIS.

A collection of zooplankton from the ocean's surface might contain a great variety of small animals. Some of them can be seen only with a microscope because they consist of only one cell. Others are a little larger, such as tiny shrimp and other crustaceans, and very

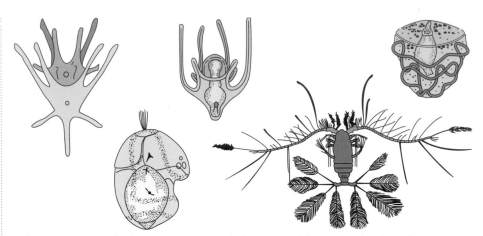

◆ The ocean's zooplankton community includes many different kinds of small marine animals. These are zooplankton organisms of several sizes and shapes.

small jellyfish. The small eggs and larvae of many kinds of FISH and other animals also drift in the ocean as zooplankton.

Zooplankton tend to stay near the surface of the ocean, where tiny marine ALGAE called phytoplankton grow in the sunlight. Phytoplankton are eaten by many animals of the zooplankton. Thus, even though phytoplankton and zooplankton together are often just called PLANKTON, they are really as different from each other as grasses and grasshoppers are.

## NO ZOOPLANKTON, NO FISHING

Plankton are the first two links in the FOOD CHAINS of the open ocean. SOLAR ENERGY is captured by phytoplankton and used to make food; zooplankton eat the phytoplankton, and other marine animals eat the zooplankton. Some food chains in the ocean are very short. For example, right WHALES and bowhead whales simply eat the zooplankton itself, by straining it from the water in their huge jaws. In other cases,

the food chain is longer. Zooplankton are eaten by small fish, which are eaten by larger fish, which are then eaten by a large PREDATOR such as a shark, DOLPHIN/PORPOISE, TUNA, seal, or human. Thus, any predator that hunts in the sea depends upon zooplankton either directly, as the bowhead whale does, or indirectly, as we do.

◆ Zooplankton feed on phytoplankton and are near the base of the ocean food chain. Here, the young stages of crabs can be seen.

# Major Environmental Events and Legislation in the United States

1813   John Audubon sees more than 1 billion passenger pigeons in Kentucky.

1827–   *The Birds of America,* by John James
1838   Audubon, is published.

1845   *Walden, or Life in the Woods,* by Henry David Thoreau, is published.

1849   The Department of the Interior is established.

1862   The Homestead Act gives land to settlers for farming. This encourages settlement of the Great Plains states, which consequently disrupts prairie ecosystems.

1864   The first state park in the United States is created at Yosemite. It will later become a national park.

1872   The Mining Act allows prospectors to stake claims on public land. If prospectors discover valuable minerals on a claim, they obtain ownership of the land.

1872   Yellowstone National Park, the first national park in the world, is established.

1880   Yosemite National Park Act established the Yosemite Valley of California as a national park.

1891   The Forest Reserve Act gives the president the power to set aside federal lands as forest reserves. The reserves are later called National Forests.

1897   The Forest Management Act permits timber cutting on forest reserves.

1899   The Rivers and Harbors Appropriations Act makes it illegal to dump waste into waters used by boats or ships, except by special permission.

1900   The Lacey Act makes the interstate shipment of meat from animals killed in violation of state laws a federal offense.

1902   The Reclamation Act provides money for dams and irrigation in the dry parts of the western United States.

1906   After many years of slaughter, only about 300 American buffalo remain of the 30 to 60 million that were seen by the first European settlers.

1908   Chlorine is first used in water treatment plants.

1914   After years of unregulated hunting of passenger pigeons, the last one dies in the Cincinnati Zoo.

1916   The National Park System is formed.

1918   The Migratory Bird Treaty Act was established in 1918 between Canada and the United States and later extended to other countries to limit the hunting, capture, selling, and killing of migratory birds, such as ducks, geese, and swans.

1920   The Mineral Leasing Act regulates mining on federal lands and allows oil and gas leasing on unreserved public lands.

1930   A severe drought ruins much of the southern Great Plains, which became known as the Dust Bowl.

1933   The Tennessee Valley Authority Act creates an organization to oversee the use and conservation of the natural resources of the Tennessee River Basin.

1934   The Taylor Grazing Act regulates grazing on federal grazing lands, the last major type of public land to be supervised by the government.

1935  The Soil Conservation Service is established to help prevent soil erosion after the drought that resulted in the Dust Bowl.

1938  The Food, Drug, and Cosmetic Act sets federal standards for the labeling of foods, drugs, and cosmetics.

1941  The number of whooping cranes dwindles to fifteen, mostly as a result of habitat destruction.

1943  A highway connecting Alaska to the mainland United States is completed.

1946  The Bureau of Land Management is established.

1947  The Federal Insecticide, Fungicide, and Rodenticide Act requires that pesticides be registered with the U.S. Department of Agriculture.

1947  *The Everglades: River of Grass,* by Marjorie Stoneman Douglas, is published.

1948  The Water Pollution Control Act gives state and local governments funds to combat water pollution.

1952  Oregon becomes the first state to have an air pollution control program.

1954  The Atomic Energy Act regulates the development of the nuclear power industry and protects the public from radioactivity.

1955  The Air Pollution Control Act gives funds to states to help prevent air pollution.

1956  Environmentalists prevent the construction of Echo Park Dam on the Colorado River.

1956  The Federal Water Pollution Control Act increases funding to states to combat water pollution and establishes standards for sewage treatment plants.

1958  Levels of carbon dioxide in the air are first measured at the Mauna Loa Research Station in Hawaii. The first level was 314 parts per million.

1960  The Hazardous Substances Act requires warning labels for dangerous household substances.

1960  The Multiple Use-Sustained Yield Act requires that the management of national forests be balanced between economic and ecological interests.

1962  *Silent Spring,* by Rachel Carson, is published.

1963  The Clean Air Act provides for federal control of local, state, and national efforts to prevent air pollution.

1964  The Wilderness Act creates a system of wilderness preserves.

1965  The Solid Waste Disposal Act is the first legislation governing waste disposal. It is later amended by the Resource Recovery Act and the Resource Conservation and Recovery Act.

1965  The Motor Vehicle Air Pollution Control Act establishes limits for automobile emissions.

1968  The Wild and Scenic Rivers Act provides for the conservation of wild and scenic rivers.

1968  The National Trails System Act promotes the development of trails that are important for their history or scenery.

1969  The National Environmental Policy Act expresses concern about the environment. It requires environmental impact statements for all major federal actions that could significantly affect the environment.

1969  The Endangered Species Conservation Act, also known as the Endangered Species Act, establishes a policy of identifying threatened and endangered species.

1970  The Environmental Protection Agency is established.

1970  The first Earth Day is celebrated on April 22.

1970    The Amendment to the Clean Air Act strengthens the 1963 law by setting a timetable for automobile emission controls, establishing air quality standards for six major pollutants, and phasing out the lead in fuels.

1970    The Resource Recovery Act amends the Solid Waste Disposal Act of 1965 by providing money for recycling programs and requiring an investigation of hazardous-waste management.

1972    The Clean Water Act amends the Federal Water Pollution Act of 1956. It states that all surface waters in the nation should be "fishable and swimmable" by 1983 and sets the goal of 1985 for zero discharge of water pollutants.

1972    The Marine Mammal Protection Act makes it illegal to import, kill, and harass marine mammals.

1972    The Marine Protection, Research, and Sanctuaries Act, also known as the Ocean Dumping Act, regulates ocean dumping and sets up a process for establishing marine sanctuaries.

1972    The Federal Insecticide, Fungicide, and Rodenticide Act replaces and strengthens the law passed in 1947. It gives the Environmental Protection Agency the power to ban or restrict the use of dangerous pesticides.

1973    The Convention of International Trade in Endangered Species of Fauna and Flora (CITES) is signed by 80 nations to prohibit the trafficking and exploitation of endangered species of plants and animals.

1973    The pesticide DDT is banned.

1973    OPEC countries stop shipment of oil to the United States.

1974    Construction begins on the Alaska Pipeline, which will bring oil from northern Alaska to the lower 48 states.

1974    The Forest and Rangeland Renewable Resources Planning Act is enacted to facilitate management of natural resources in the National Park System.

1974    The Safe Drinking Water Act gives the Environmental Protection Agency the power to set water quality standards and to monitor water supplies.

1975    The Energy Policy and Conservation Act sets standards to decrease fuel usage in new automobiles.

1975    The Hazardous Materials Transportation Act establishes minimum standards for the transportation of hazardous materials.

1975    Atlantic salmon are seen in the Connecticut River after a 100-year absence.

1976    The Resources Conservation and Recovery Act amends the Solid Waste Disposal Act of 1965. It requires the Environmental Protection Agency to track hazardous wastes from "cradle to grave" and regulates the transportation and treatment of hazardous wastes.

1976    The Toxic Substances Control Act gives the Environmental Protection Agency the power to regulate hazardous chemicals.

1977    Amendments to the Clean Air Act of 1963 set up new standards and procedures, including a requirement to use the "best available control technology" (BACT). They also establish programs to protect the air quality in unpolluted areas.

1977    The Surface Mining Control and Reclamation Act requires that surface mines be reclaimed by being filled and having their vegetation restored and their water quality repaired. This act makes it illegal to use surface mining in national parks, forests, and other designated areas.

1978    The National Energy Conservation Policy Act (called the "gas guzzler tax") imposes a tax on automobiles that use fuel inefficiency.

This act provides tax credits for residences that conserve energy and use solar energy. It also requires utilities to offer conservation programs to customers and requires that the energy efficiency of federal buildings be improved. Grants are provided for energy conservation in homes and in public buildings. The price of natural gas is increased, partly through the elimination of federal price regulations. This act calls for industry to convert to coal as an energy source.

1978 The Outer Continental Shelf Lands Act states that industry is accountable for oil pollution on the outer continental shelf.

1978 The Quiet Communities Act amends the Noise Control Act. It establishes standards for noise levels in aircraft and products and imposes penalties for noise pollution.

1978 Love Canal, Niagara Falls, is evacuated after it is discovered that the residential area was once a chemical-waste dump.

1979 The Three Mile Island nuclear reactor comes close to a nuclear meltdown. No new nuclear power plants are built after this accident.

1980 The Acid Precipitation Act calls for research on the causes and effects of acid precipitation.

1980 The Alaska National Interest Lands Conservation Act establishes 56 million acres of parks, wildlife refuges, and wilderness areas in Alaska.

1980 The Comprehensive Environmental Response, Compensation, and Liability Act, or Superfund, establishes a $1.6 billion fund to pay for the cleanup of improperly disposed hazardous waste. It authorizes the Environmental Protection Agency to demand repayment from the polluters.

1980 The Asbestos School Hazard Detection and Control Act establishes a program to reduce the risks from asbestos in school buildings.

1980 The Low Level Radioactive Waste Policy Act requires states to be responsible for the disposal of the low-level radioactive waste they produce.

1982 The Nuclear Waste Policy Act requires the Department of Energy to develop a permanent site for the storage of high-level nuclear waste by 1998.

1984 Amendments to the Resource Conservation and Recovery Act expand the role of the federal government in regulating waste disposal; establish new requirements for disposal facilities; and encourage states to establish recycling programs for city waste.

1985 *Nature* publishes the first report of ozone thinning in the atmosphere above Antarctica.

1985 The Food Security Act encourages farmers to develop plans to prevent erosion, while discouraging them from converting wetlands into agricultural land.

1986 The Superfund Amendments and Reauthorization Act increases the money available for the cleanup of hazardous waste sites to $8.5 billion. It requires hazardous waste facilities to report releases of toxic chemical wastes and requires emergency planning for hazardous waste crises.

1986 Times Beach, Missouri, is evacuated after high levels of dioxin are detected.

1987 The Marine Plastic Pollution Research and Control Act makes it illegal to dump plastic into the ocean. This act also restricts the dumping of other wastes into the ocean and requires all ports to have garbage-disposal facilities for ships.

1987 The National Appliance Energy Conservation Act requires that appliances use 25% less energy by 1993.

1987    The Water Quality Act amends the Clean Water Act of 1972. It requires states to control nonpoint source pollution and to reduce toxic pollutants.

1987    A barge carrying garbage from a town in New York sails up and down the coast for months in search of a disposal site.

1987    Yucca Mountain in Nevada is designated as the permanent storage site for high-level nuclear waste.

1988    The Ocean Dumping Reform Act makes it illegal to dump sewage sludge after 1991.

1988    The Alternative Motor Fuels Act encourages the development of automobiles that use alternative fuels such as ethanol and methanol.

1988    The Medical Waste Tracking Act creates a tracking system for medical waste.

1988    The Safe Drinking Water Act Amendment makes it illegal to use lead pipes in city water systems.

1989    The North American Wetlands Protection Act provides money for the protection and restoration of wetlands.

1989    The oil supertanker *Exxon Valdez* runs aground in Prince William Sound, Alaska, spilling 11 million gallons of oil into the water.

1990    The number of insect species that have become resistant to pesticides reaches 500.

1990    The Oil Pollution Act is passed in response to the *Exxon Valdez* disaster. It requires that oil companies pay for cleanup of oil spills, that oil tankers and barges have double hulls by 2015, and that oil tankers have trained crews. This act also establishes a fund for cleanup of spills when the polluter is unknown.

1990    The Pollution Prevention Act establishes a new office in the Environmental Protection Agency designed to help industries limit pollutants.

1990    Amendments to the Clean Air Act provide for a 50% reduction in sulfur dioxide and nitrogen oxides to reduce acid rain and a 70% reduction in carbon monoxide from automobiles.

1990    The National Environmental Education Act establishes an Office of Environmental Education in the Environmental Protection Agency, which provides programs, grants, and internships for environmental education.

1992    The Earth Summit is held in Rio de Janeiro, Brazil.

1992    The Marine Mammal Protection Act Amendment authorizes the Secretary of State to enter into international agreements to prevent the use of tuna-fishing methods that harm dolphins or other marine mammals.

1992    The Wild Bird Conservation Act makes it illegal to import exotic birds beginning in 1996.

1994    The level of carbon dioxide in the air, measured at the Mauna Loa Research Station, reaches 358 parts per million.

1995    As a result of habitat protection and captive breeding, the number of whooping cranes increases to 230.

1995    Protected by law, American buffalos, which numbered 300 in 1906, now number more than 130,000.

# Environmental Organizations

Adopt-a-Stream Foundation
P.O. Box 5558
Everett, WA 98206
(206) 316-8592

*Promotes stream restoration with the aim of having every stream adopted by nearby residents. Has three publications:* Adopting a Stream, Adopting a Wetland, *and* Stream Keepers Field Guide, *as well as a video entitled* The Stream Keeper.

African Wildlife Foundation
1717 Massachusetts Avenue NW
Washington, DC 20036
(202) 265-8394

*Works with Africans to establish and support conservation education programs.*

Air and Waste Management Association
1 Gateway Center, 3rd Floor
Pittsburgh, PA 15222
(412) 232-3444

*Organization for environment professionals. Supports environmental education at all levels.*

Alaska Wildlife Alliance
Box 202022-H
Anchorage, AK 99520
(907) 277-0897

*Publicizes wildlife issues in Alaska. Recent concerns include the protection of wolves and humpback whales.*

Alliance for Environmental Education
9309 Center Street, Suite 101
Manassas, VA 22110
(703) 330-5667

*Sponsors educational programs in colleges, universities, and other institutions.*

Alliance to Save Energy
1725 K Street NW, Suite 509
Washington, DC 20006
(202) 857-0666

*Coalition of environmental, governmental, and business leaders dedicated to conserving energy.*

Alternative Energy Resources Organization
25 South Ewing, Room 214
Helena, MT 59601
(406) 443-7272

*Focuses on energy conservation and renewable resources in the Great Plains and northern Rocky Mountains.*

American Cave Conservation Association
American Cave and Karst Center
P.O. Box 409
Horse Cave, KY 42749
(502) 786-1466

*Promotes the conservation of caves and cave resources. Focuses on the prevention of ground-water pollution in Karst areas. Offers workshops, maintains a clearinghouse of information, and operates a museum and educational center.*

American Council for an Energy-Efficient Economy
1001 Connecticut Avenue NW, Suite 801
Washington, DC 20036
(202) 429-8873

*Promotes energy-efficient technology and policy through research.*

American Forest and Paper Association
1111 19th Street NW, Suite 800
Washington, DC 20036
(202) 463-2462

*Sponsors Project Learning Tree, an educational program for young people. Provides materials to teachers.*

American Forests
Global ReLeaf Program
P.O. Box 2000
Washington, DC 20013
(202) 667-3300

*Supports international tree-planting and educational programs.*

American Geographical Society
156 Fifth Avenue, Suite 600
New York, NY 10010
(212) 242-0214

*Provides travel programs, awards, and consulting services to businesses and schools. Emphasizes world issues. Publishes* The Geographical Review *and* Focus *magazines.*

American Hiking Society
P.O. Box 20160
Washington, DC 20041
(301) 565-6704

*Helps preserve hiking trails through education and a public information service. Developed the first coast-to-coast hiking trail in the United States.*

American Littoral Society
Sandy Hook
Highlands, NJ 07732
(908) 291-0055

*Champions the protection of marine and coastal habitats. Publishes the* Underwater Naturalist *and a national newsletter.*

American Nature Study Society
5881 Cold Brook Road
Homer, NY 13077
(607) 749-3655

*Conducts workshops for environmental educators and presents awards for published works that reflect the excitement of discovery and hands-on experience.*

American Nuclear Society
555 North Kensington Avenue
LaGrange Park, IL 60525
(708) 579-8265

*Promotes the understanding and advancement of nuclear science and technology.*

American Rivers
1025 Vermont Avenue NW, Suite 720
Washington, DC 20005
(202) 547-6900

*Promotes the preservation and restoration of America's system of rivers. Emphasizes clean water, fish and wildlife, recreation, and beauty.*

American Society of Agricultural Engineers
2950 Niles Road
St. Joseph, MI 49085
(616) 429-0300

*Professional organization dedicated to the advancement of agricultural engineering, as well as to the protection of the environment and natural resources.*

American Society of Agronomy
Crop Science Society of America
Soil Science Society of America
677 South Segoe Road
Madison, WI 53711
(608) 273-8080

*Works together to conserve natural resources to produce food crops while also protecting the environment.*

American Society for Environmental History
Center for Technology Studies
New Jersey Institute of Technology
University Heights
Newark, NJ 07012
(201) 596-3334

*Focuses on research, publishing, and teaching related to environmental history.*

Americans for the Environment
1400 16th Street NW, Box 24
Washington, DC 20036
(202) 797-6665

*Provides training for environmental activists in political campaigns and elections.*

Antarctica Project
P.O. Box 76920
Washington, DC 20013
(202) 544-0236

*Educates the public about Antarctica and reports on governmental and scientific meetings.*

Asbestos Information Association
1745 Jefferson Davis Highway, Suite 406
Arlington, VA 22202
(703) 412-1150

*Provides information to the public about asbestos.*

Association of Foam Packaging Recyclers
1275 K Street NW, Suite 400
Washington, DC 20005
(800) 944-8448; (202) 371-2491

*Encourages recycling of foam packaging and the use of recycled foam products.*

Biomass Energy Research Association
1825 K Street NW, Suite 503
Washington, DC 20006
(800) 247-1755

*Supports research and information exchange about energy from biomass.*

Caretta Research Project
Savannah Science Museum
4405 Paulsen Street
Savannah, GA 31405
(912) 355-6705

*Studies and protects the threatened loggerhead sea turtle on Warsaw Island National Wildlife Refuge. Open to volunteers who may wish to participate in the research.*

Center for Alaskan Coastal Studies
P.O. Box 2225
Homer, AK 99603
(907) 235-6667

*Provides educational programs about marine and coastal environments to Alaskan schoolchildren and the public.*

Center for Environmental Information
50 West Main Street
Rochester, NY 14614
(716) 262-2870

*Provides information about environmental issues, including ethics, laws, and education.*

Center for Marine Conservation
1725 DeSales Street NW, Suite 500
Washington, DC 20036
(202) 429-5609

*Focuses on the conservation of marine habitats and wildlife and on the prevention of marine pollution.*

Center for Plant Conservation
Missouri Botanical Garden
P.O. Box 299
St. Louis, MO 63166
(314) 577-9450

*Conserves rare and endangered plants through cultivation at botanical gardens and provides information to the public.*

Citizens for a United Earth
1880 Route 64
Ionia, NY 14475
(716) 624-3673

*Writes letters to protest environmental destruction, overpopulation, and war. Publishes a monthly newsletter.*

Clear Water Action
1320 18th Street NW, Suite 300
Washington, DC 20036
(202) 457-1286

*Works for clean water, safe disposal of solid waste and toxic chemicals, and protection of wetlands and coastal waters.*

Coalition for Environmentally Responsible
  Economies
711 Atlantic Avenue, 4th Floor
Boston, MA 02111
(617) 451-9495

Organization of environmental groups, public interest groups, social investors, and labor groups. Promotes environmentally responsible economic growth.

Colorado Environmental Coalition
777 Grant Street, Suite 606
Denver, CO 80203
(303) 837-8701

*Promotes the protection of natural resources, especially those in the state of Colorado.*

Conservation International
1015 18th Street NW, Suite 1000
Washington, DC 20036
(202) 429-5660

*Works to save endangered species and ecosystems. Gives financial and technical assistance to local communities worldwide to help preserve their biological resources.*

Council on Economic Priorities
30 Irving Place
New York, NY 10003
(212) 420-1133

*Researches environmental and social-responsibility issues. Publishes two guides,* Shopping for a Better World *and* Students Shopping for a Better World.

Cousteau Society, Inc.
870 Greenbrier Circle, Suite 402
Chesapeake, VA 23320
(804) 523-9335

*Produces films, television programs, books, and magazine articles dedicated to the protection of all living things. Sponsors lectures and field-study programs.*

Defenders of Wildlife
1244 19th Street NW
Washington, DC 20036
(202) 659-9510

*Protects wildlife and habitats through education and litigation.*

Delta Waterfowl Foundation
P.O. Box 3128
Bismarck, ND 58502
(701) 222-8857

*Promotes the conservation of wildlife, soil, water, and wetlands.*

Ducks Unlimited
1 Waterfowl Way
Memphis, TN 38120
(901) 758-3825

*Promotes the conservation and restoration of North American wetlands.*

Earth Island Institute
300 Broadway, Suite 28
San Francisco, CA 94133
(415) 788-3666

*Helps develop projects for the conservation and restoration of the global environment. Projects include the International Marine Mammal Project, the Sea Turtle Restoration Project, International Green Circle, the Urban Habitat Program, and Baikal Watch.*

Earthtrust
25 Kaneohe Bay Drive, Suite 205
Kailua, HI 96734
(808) 254-2866

*Promotes wildlife conservation and the protection of endangered species, including whales, dolphins, rhinos, and tigers. Emphasizes long-term political solutions.*

Earthwatch
680 Mount Auburn Street
P.O. Box 403N
Watertown, MA 02272
(617) 926-8200

*Sponsors scholarly field research by arranging for paying volunteers to assist scientists on expeditions around the world. Concentrates on endangered habitats and species, world health, and international cooperation.*

Ecology Center
2530 San Pablo Avenue
Berkeley, CA 94702
(510) 548-2220

*Aims to demonstrate workable alternatives to environmentally destructive practices. Operates a library and a mail-order bookstore. Publishes a monthly newsletter.*

Energy Research Institute
6850 Rattlesnake Hammock Road
Naples, FL 33962
(941) 793-1922

*Focuses on learning how to harness alternative energy sources such as solar and wind energy.*

Environmental Action
6930 Carroll Avenue, Suite 600
Takoma Park, MD 20912
(301) 891-1100

*Encourages the prevention of pollution and the conservation of natural resources.*

Environmental Defense Fund
257 Park Avenue South, 16th Floor
New York, NY 10010
(212) 505-2100

*Links law, science, and economics to create innovative solutions to worldwide environmental problems.*

Environmental Media Corporation
P.O. Box 1016
Chapel Hill, NC 27514
(919) 933-3003

*Produces media for environmental education in schools and homes.*

Foundation for Field Research
P.O. Box 2010
Alpine, CA 91903
(619) 445-9264

*Arranges for paid volunteers to assist scientists in the field. Areas of research include ecology, conservation, anthropology, and archaeology.*

Friends of the Earth
1025 Vermont Avenue NW, Suite 300
Washington, DC 20005
(202) 783-7400

*Lobbies Congress and disseminates public information about conservation, environmental protection, and rational use of resources.*

Friends of the Sea Otter
2150 Garden Road, Suite B4
Monterey, CA 93940
(408) 373-2747

*Works to protect the sea otter and its habitat.*

Fund for Animals
200 West 57th Street
New York, NY 10019
(212) 246-2096

*Opposes cruelty to animals, both wild and domestic, and supports the preservation of biodiversity.*

Greenpeace USA
1436 U Street NW
Washington, DC 20009
(202) 462-1177

*Campaigns for protection of the environment. Campaigns include the protection of marine ecology and the elimination of nuclear pollution, toxic pollution, and atmospheric destruction.*

Hawaii Nature Center
2131 Makiki Heights Drive
Honolulu, HI 96822
(808) 955-0100

*Fosters awareness, appreciation, and understanding of the Hawaiian islands through field experiences for elementary schoolchildren and the public. Sponsors hands-on field experiences at field sites on the islands of A'ahu and Maui.*

Hawk-Watch International, Inc.
P.O. Box 35706
Albuquerque, NM 87176
(505) 255-7622

Through education and research, supports the conservation of birds of prey and their habitats in the western United States.

Household Hazardous Waste Project
1031 East Battlefield, Suite 214
Springfield, MO 65807
(417) 889-5000

*Provides training, educational materials, consulting, and a referral and information service. Participates in local, regional, state, and national meetings that address health and environmental concerns related to household hazardous products and wastes.*

Inform
120 Wall Street, 16th Floor
New York, NY 10016
(212) 361-2400

*Through research and education, identifies and reports on practical actions for the preservation and conservation of natural resources and public health.*

Institute for Earth Education
P.O. Box 115
Greenville, WV 24945
(304) 832-6404

*Develops and disseminates educational materials, including a journal, conferences, and catalog of books.*

Institute for Environmental Education
18554 Haskins Road
Chagrin Falls, OH 44023
(718) 543-7303

*Publishes and sells environmental education materials.*

International Alliance for Sustainable Agriculture
1701 University Avenue SE
Minneapolis, MN 55414
(612) 331-1099

*Works for the worldwide realization of sustainable agriculture—food systems that are ecologically sound, economically viable, and socially just and humane.*

International Bird Rescue Research Center
699 Potter Street
Berkeley, CA 94710
(510) 841-9086

*Provides and maintains rapid response capabilities for oil spills involving wildlife. Provides training, consultation, and planning to promote preparedness for response to oil spills.*

Izaak Walton League of America
Save Our Streams
707 Conservation Lane
Gaithersburg, MD 20878
(301) 548-0150

*Encourages interested individuals and groups to monitor streams.*

LaMotte Company
P.O. Box 329
Route 213 North
Chestertown, MD 21620
(410) 778-3100

*Publishes* The Monitor's Handbook, *a reference booklet on water analysis; develops and manufactures self-contained, portable test kits and electronic instruments designed for water, soil, and air analysis.*

Land and Water Fund of the Rockies
2260 Baseline, Suite 200
Boulder, CO 80302
(303) 444-1188

*Provides free legal aid to grassroots environmental groups in Arizona, Colorado, Idaho, Montana, New Mexico, Utah, and Wyoming.*

League of Conservation Voters
1707 L Street NW, Suite 750
Washington, DC 20036
(202) 785-8683

*Publishes a national environmental score card that reflects Congress's voting on key environmental issues.*

Marine Mammal Center
Marin Headlands, Golden Gate National
 Recreation Area
Sausalito, CA 94965
(415) 289-7325

*Rescues, rehabilitates, and studies marine mammals. Provides information through science and education programs.*

Marine Mammal Stranding Center
P.O. Box 773
3625 Brigantine Boulevard
Brigantine, NJ 08203
(609) 266-0538

*Rescues and rehabilitates marine mammals and sea turtles that are found stranded in New Jersey and surrounding states.*

National Arbor Day Foundation
100 Arbor Avenue
Nebraska City, NE 68410
(402) 474-5655

*World's largest tree-planting environmental organization.*

National Association of Noise Control Officials
53 Cubberly Road
Trenton, NJ 08690
(609) 586-2684

*Provides opportunities for information exchange, discussion, and cooperative study. Promotes laws to control noise pollution. Cooperates with industry and the scientific community to reduce excessive and unnecessary noise.*

National Audubon Society
700 Broadway
New York, NY 10003
(212) 979-3000

*Grassroots environmental organization concerned with protecting and preserving ancient forests,*

*wetlands, endangered species, the Arctic National Wildlife Refuge, the Platte River, the Everglades, and the Adirondack Park.*

National Coalition Against the Misuse of Pesticides
701 E Street SE, Suite 200
Washington, DC 20003
(202) 543-5450

*Coalition of health, environmental, labor, farm, consumer, and church groups. Maintains a practical information hotline on toxic hazards and non-chemical pest control.*

National Energy Foundation
5225 Wiley Post Way, Suite 170
Salt Lake City, UT 84116
(801) 539-1406

*Creates and distributes instructional materials to teachers. Materials deal with energy, water, mineral resources, science, technology, the environment, and other natural resource topics. Provides kits, lesson plans, activity books, and posters.*

National Food and Energy Council
409 Vandiver Drive, Suite 4-202
Columbia, MO 65202
(314) 875-7155

*Develops materials for educational institutions and electric cooperatives for efficient electric and energy use.*

National Solid Wastes Management Association
1730 Rhode Island Avenue NW, Suite 1000
Washington, DC 20036
(202) 659-0708

*Represents more than 2,500 North American companies involved in the collection, recycling, treatment, and disposal of solid, hazardous, and medical wastes.*

National Wildflower Research Center
4801 La Crosse Avenue
Austin, TX 78739
(512) 292-4200

Promotes the preservation and reestablishment of native wildflowers, grasses, shrubs, and trees in North America.

National Wildlife Federation
P.O. Box 8925
Vienna, VA 22183
(800) 822-9919

Distributes educational materials. Sponsors outdoor nature programs. Lobbies Congress in matters of environmental disputes in an effort to conserve fisheries, wildlife, and natural resources.

National Wildlife Refuge Association
P.O. Box 60318
Potomac, MD 20859
(301) 983-9498

Advocacy organization representing wildlife professionals and concerned citizens.

Nature Conservancy
1815 North Lynn Street
Arlington, VA 22209
(703) 841-5300

Committed to protecting plants, animals, and natural communities.

New Alternatives Fund, Inc.
150 Broadhollow Road
Melville, NY 11747
(516) 423-7373

A mutual fund that concentrates on investments in solar energy, geothermal energy, natural gas, energy efficient devices, and energy conservation.

Nuclear Information and Resource Service
1424 16th Street NW, Suite 404
Washington, DC 20036
(202) 328-0002

Maintains networking and information clearinghouse for environmental activists concerned with nuclear power and waste issues.

Pacific Whale Foundation
101 North Kihei Road
Kihei, HI 96753
(808) 879-8860/1-800-WHALE 11

Develops marine mammal education materials, public awareness activities, and educational marine recreation programs, such as Adopt-a-Whale; free whale watches for Maui schoolchildren, school visits, and internships. Lobbies for more effective environmental protection.

PKI Kettle Moraine Division
604 2nd Avenue
West Bend, WI 53095
(414) 334-4978

Develops and implements projects and programs that promote combined environmental education and economic and humanitarian benefit. Emphasizes youth programs.

Polystyrene Packaging Council, Inc.
1025 Connecticut Avenue NW, Suite 515
Washington, DC 20036
(202) 822-6424

Represents the major producers of polystyrene food service products and the suppliers of resin. Works with local governments and the public to increase awareness of polystyrene recycling.

Population Action International
1120 19th Street NW, Suite 550
Washington, DC 20036
(202) 659-1833

Advocates universal voluntary access to family planning to achieve world population stability.

Public Citizen
1600 20th Street NW
Washington, DC 20009
(202) 588-1000

Represents consumer interests through lobbying, litigation, research, and publications.

Rachel Carson Council, Inc.
8940 Jones Mill Road
Chevy Chase, MD 20815
(301) 652-1877

*Acts as an international clearinghouse of information on ecology of the environment for scientists and the general public.*

Rails-to-Trails Conservancy
1400 16th Street NW, Suite 300
Washington, DC 20036
(202) 797-5400

*Converts old railroad corridors into public trails connecting major cities for hiking, bicycling, horseback riding, and cross-country skiing.*

Rainforest Action Network
450 Sansome Street, Suite 700
San Francisco, CA 94111
(415) 398-4404

*Activist organization concerned with United States tropical timber imports, multilateral development banks in the rain forest, rights of indigenous peoples, Japanese corporations in the rain forest, Hawaiian rain forests, oil exploration, Amazonia, New Guinea, Sarawak, Malaysia, and cattle ranching.*

Rainforest Alliance
65 Bleecker Street
New York, NY 10012
(212) 677-1900

*Promotes sound alternatives to activities that cause tropical deforestation.*

Renew America
1400 16th Street NW, Suite 710
Washington, DC 20036
(202) 232-2252

*Distributes the annual* Environmental Success Index *to policy makers, business leaders, industry, government, and the media.*

Rocky Mountain Institute
1739 Snowmass Creek Road
Snowmass, Colorado 81654
(970) 927-3851

*Nonprofit resource policy foundation that focuses on water and agriculture programs.*

Ruffed Grouse Society
451 McCormick Road
Coraopolis, PA 15108
(412) 262-4044

*Promotes the improvement of the environment for the ruffed grouse. Provides direct financial assistance to public land managers and educational information to private landowners.*

Sierra Club
730 Polk Street
San Francisco, CA 94109
(415) 776-2211

*Promotes responsible use of the earth's ecosystems and resources. Current projects include old-growth forest protection, global warming, auto fuel efficiency, energy policy, toxic-waste regulations, tropical-forest preservation, and protection of wildlife.*

Smithsonian Institution Office of Environmental Awareness
S. Dillon Ripley Center, Suite 3123
MRC 705
Washington, DC 20560
(202) 357-4797

*Gathers and disseminates information about a wide range of environmental issues. Works with Smithsonian bureaus and outside groups to reach the general public and professional audiences through exhibitions, publications, conferences, and workshops.*

Soil and Water Conservation Society
7515 Northeast Ankeny Road
Ankeny, IA 50021
(515) 289-2331

*Advocates the conservation of soil, water, and related natural resources.*

Southwest Research and Information Center
P.O. Box 4524
Albuquerque, NM 87106
(505) 262-1862

*Provides information to the public on environmental matters, human health, and communities.*

Steel Recycling Institute
680 Andersen Drive
Pittsburgh, PA 15220
(412) 922-2772

*Provides primary information and acts as a technical resource for recyclers, municipalities, legislators, educators, businesses, and other entities.*

Student Conservation Association, Inc.
P.O. Box 550
Charlestown, NH 03603
(603) 543-1700

*Provides educational opportunities for volunteers to assist with the stewardship of public lands and natural resources.*

Trout Unlimited
1500 Wilson Boulevard, Suite 310
Arlington, VA 22209
(703) 522-0200

*Provides information to legislators on legislative and regulatory issues pertaining to cold-water fisheries.*

Union of Concerned Scientists
2 Brattle Square
P.O. Box 9105
Cambridge, MA 02238
(617) 547-5552

*Conducts technical studies and public education regarding environmental and security threats facing humanity. Seeks to influence government policy at local, state, federal, and international levels.*

Nuclear Energy Institute
1776 I Street NW, Suite 400
Washington, DC 20006
(202) 739-8000

*National public relations and communications arm of the nuclear energy industry.*

Urban Options
405 Grove Street
East Lansing, MI 48823
(517) 337-0422

*Provides educational materials and programs to help improve the environmental quality of urban areas.*

Washington Citizens for Recycling
157 Yesler Way, Suite 309
Seattle, WA 98104
(206) 343-5171

*Promotes recycling and resource conservation through public education, the lobbying of citizens, and by working with government and industries.*

Water Environment Federation
601 Wythe Street
Alexandria, VA 22314
(800) 666-0206

*Organization of water quality specialists from around the world.*

Wild Canid Survival and Research Center
P.O. Box 760
Eureka, MO 63025
(314) 938-5900

*Dedicated to preserving the wolf and its place in the natural ecosystem by reintroducing animals into the wild. Provides for scientific observation of wolf behavior while ensuring adequate nutrition and veterinary care.*

Wildlife Conservation Society
185th Street and Southern Boulevard
Bronx, NY 10460
(718) 220-6891

*Works with local governments in more than 40 countries to gather ecosystem data. Helps to devise plans for protecting endangered wildlife and habitats.*

Wildlife Society
5410 Grosvenor Lane
Bethesda, MD 20814
(301) 897-9770

*Conserves and sustains wildlife productivity and diversity through resource management. Promotes the enhancement of the scientific and technical capability and performance of wildlife professionals.*

World Resources Institute
1709 New York Avenue NW, Suite 700
Washington, DC 20006
(202) 638-6300

*Provides policy research and technical assistance to governments, the private sector, environmental and development organizations, and other groups.*

World Society for the Protection of Animals
P.O. Box 190
Boston, MA 02130
(617) 522-7000

*Helps animals affected by both natural and human-made disasters.*

Worldwatch Institute
1776 Massachusetts Avenue NW
Washington, DC 20036
(202) 452-1999

*Informs policy makers, the press, and the public about the complex links between the global economy and the environment on which it is based.*

World Wildlife Fund
1250 24th Street NW, Suite 400
Washington, DC 20037
(202) 293-4800

*Works worldwide to preserve endangered wildlife and wildlands by encouraging sustainable development and the preservation of biodiversity.*

Zero Population Growth, Inc.
1400 16th Street NW, Suite 320
Washington, DC 20036
(202) 332-2200

*Works to bring about a balance between population, resources, and the environment. Publishes a newsletter, population fact sheets, the 177-page book* Earth Matters, *and a catalog of teaching materials on population. Conducts teachers' workshops on population education.*

# United States Government Organizations with Environmental Responsibility

Army Corps of Engineers
Department of Defense
20 Massachusetts Avenue NW
Washington, DC 20314
(202) 272-0010

*Responsible for protecting wetlands and tidal zones and overseeing the construction of bridges, dams, and reservoirs.*

Bureau of Land Management
Department of the Interior
Main Interior Building
18th and C Streets NW
Washington, DC 20240
(202) 343-5717

*Manages mineral resources and public lands.*

Bureau of Reclamation
Department of the Interior
Main Interior Building
18th and C Street NW
Washington, DC 20240
(202) 477-8732

*Operates federal energy and water programs, mainly in the western United States.*

Coast Guard
Department of Transportation
2100 2nd Street SW
Washington, DC 20593
(202) 267-2229

*Oversees cleanup operations after oil spills. Responsible for inspecting international vessels for hazardous materials.*

Council on Environmental Quality
Executive Office of the President
722 Jackson Place
Washington, DC 20506
(202) 395-5750

*Coordinates compliance with environmental laws under the National Environmental Policy Act. Advises the president on environmental matters.*

Environmental Protection Agency
401 M Street SW
Washington, DC 20460
(202) 382-2080

*Performs environmental research and enforces regulations related to concerns such as pollution, waste disposal, pesticides, and nuclear radiation.*

Federal Energy Regulatory Commission
Department of Energy
815 North Capitol Street
Washington, DC 20426
(202) 357-8118

*Inspects and licenses electric utilities, hydroelectric projects, and dams.*

Fish and Wildlife Service
Department of the Interior
Main Interior Building
Washington, DC 20240
(202) 343-5634

*Responsible for fish and wildlife conservation. Enforces the Endangered Species Act.*

Forest Service
Department of Agriculture
14th Street and Independence Avenue SW
P.O. Box 96090
Washington, DC 20090
(202) 447-3957

*Manages national forests and grasslands. Regulates logging in the national forests.*

National Marine Fisheries Service
Department of Commerce
1335 East-West Highway
Silver Springs, MD 20910
(301) 427-2370

*Responsible for the protection of whales, porpoises, seals, sea lions, and other marine mammals.*

National Oceanic and Atmospheric Administration
Department of Commerce
14th Street and Constitution Avenue NW
Washington, DC 20230
(202) 377-8090

*Oversees the National Weather Service. Responsible for environmental protection of coasts and inland waterways.*

National Park Service
Department of the Interior
Main Interior Building
Washington, DC 20240
(202) 343-4747

*Manages all national parks.*

Office of Energy Efficiency and Renewable Energy
Department of Energy
1000 Independence Avenue SW
Washington, DC 20585
(202) 586-9220

*Oversees all energy-conservation programs, including research in solar, wind, geothermal, and other renewable-energy resources.*

Office of Surface Mining Reclamation and
   Enforcement
Department of the Interior
1951 Constitution Avenue NW
Washington, DC 20240
(202) 208-2953

*Regulates all strip mining and the reclamation of mined land.*

Soil Conservation Service
Department of Agriculture
14th Street and Independence Avenue SW
P.O. Box 2890
Washington, DC 20013
(202) 447-4543

*Manages all federal soil and water conservation programs.*

# Selected Bibliography

Allaby, Michael. *The Concise Oxford Dictionary of Ecology*. New York: Oxford University Press, 1994.

———, ed. *The Oxford Dictionary of Natural History*. Oxford and New York: Oxford University Press, 1986.

Anderson, Stanley H., Ronald E. Beiswenger, and P. Walton Purdom. *Environmental Science*. New York: Macmillan, 1993.

Ashworth, William. *The Late, Great Lakes*. New York: Alfred A. Knopf, 1986.

Attenborough, David. *Life on Earth: A Natural History*. Boston: Little, Brown and Company, 1979.

Barnes, Robert D. *Invertebrate Zoology*. 4th ed. Philadelphia: Saunders College Publishing, 1980.

Begon, Michael, J. L. Harper, and C. R. Townsend. *Ecology: Individuals, Populations and Communities*. 2nd ed. Boston: Blackwell Scientific Publications, 1990.

Blumberg, Louis, and Robert Gottlieb. *War on Waste*. Washington, DC: Island Press, 1989.

Bonnifield, Paul. *The Dust Bowl: Men, Dirt, Depression*. Albuquerque: University of New Mexico Press, 1979.

Botkin, Daniel B., and Edward Keller. *Environmental Science: Earth as a Living Planet*. New York: John Wiley and Sons, 1995.

Bowler, Peter J. *The Norton History of the Environmental Sciences*. New York: W. W. Norton, 1992.

Bresser, A. H., ed. *Acid Precipitation*. New York: Springer Verlag, 1990.

Brewer, Richard. *Principles of Ecology*. Philadelphia: Saunders College Publishing, 1979.

Brooks, Paul. *The House of Life: Rachel Carson At Work*. Boston: Houghton Mifflin, 1972.

Brusca, Richard C., and Gary J. Brusca. *Invertebrates*. Sunderland, MA: Sinauer Associates, 1990.

Campbell, Neil A. *Biology*. Menlo Park, CA: Benjamin/Cummings Publishing, 1987.

Caras, Roger A. *North American Mammals*. New York: Meredith Press, 1967.

Chapman, V. J. *Coastal Vegetation*. 2nd ed. New York: Pergamon Press, 1978.

Chiras, Daniel D. *Environmental Science: Action for a Sustainable Future*. 4th ed. Menlo Park, CA: Benjamin/Cummings Publishing, 1994.

———. *Environmental Science: A Framework for Decision-Making*. Menlo Park, CA: Addison-Wesley, 1989.

Clepper, Henry. *Professional Forestry in the United States*. Baltimore: Johns Hopkins Press, 1971.

Coleman, Daniel A. *Ecopolitics: Building a Green Society*. New Brunswick, NJ: Rutgers University Press, 1994.

Cousteau, Jacques-Yves, and Philippe Diole. *Life and Death in a Coral Sea*. New York: Doubleday, 1971.

———. *The Whale: Mighty Monarch of the Sea*. New York: Doubleday, 1972.

Crump, Andy. *Dictionary of Environment and Development: People, Places, Ideas and Organizations*. Cambridge, MA: The MIT Press, 1993.

Curry-Lindahl, Kai. *Conservation for Survival*. New York: William Morrow, 1972.

Curtis, Helena, and N. Sue Barnes. *Biology*. 5th ed. New York: Worth Publishers, 1983.

Dashefsky, H. Steven. *Environmental Literacy*. New York: Random House, 1993.

Dasmann, Raymond F. *Wildlife Biology*. 2nd ed. New York: John Wiley and Sons, 1981.

Davis, Charles E. *The Politics of Hazardous Waste*. Englewood Cliffs, NJ: Prentice-Hall, 1993.

Denison, Richard A., ed., *Recycling and Incineration: Evaluating the Choices*. Environmental Defense Fund, 1990.

Dixon, Bernard. *Magnificent Microbes*. New York: Atheneum, 1976.

————. *Power Unseen: How Microbes Rule the World*. New York: W. H. Freeman, 1994.

Dubos, Rene. *Health and Disease*. New York: Time-Life Books, 1965.

Duedall, Iver W., et. al., eds. *Wastes in the Ocean*. New York: John Wiley and Sons, 1985.

Ehrlich, Paul R., and Jonathan Roughgarden. *The Science of Ecology*. New York: Macmillan, 1987.

Fergusson, Jack E. *The Heavy Elements: Chemistry, Environmental Impact, and Health Effects*. New York: Pergamon Press, 1990.

Fleisher, Paul. *Ecology A to Z*. New York: Dillon Press, 1994.

Franck, Irene, and David Brownstone. *The Green Encyclopedia*. New York: Prentice-Hall, 1992.

Freedman, Bill. *Environmental Ecology: The Ecological Effects of Pollution, Disturbance, and Other Stresses*. 2nd ed. San Diego: Academic Press, 1995.

Fricke, Hans W. *The Coral Seas*. New York: G. P. Putnam's Sons, 1973.

Futuyma, Douglas J. *Evolutionary Biology*. 2nd ed. Sunderland, MA: Sinauer Associates, 1986.

George, Uwe. *In the Deserts of This Earth*. New York: Harcourt Brace Jovanovich, 1977.

Godfrey, Paul J., and Melinda M. Godfrey. *Barrier Island Ecology of Cape Lookout National Seashore and Vicinity, North Carolina*. National Park Service Scientific Monograph Series, No. 9. Washington, DC: GPO, 1976.

Gore, Jr., Albert. *Earth in the Balance*. New York: Plume, 1992.

Goudie, Andrew. *The Human Impact on the Natural Environment*. Cambridge, MA: The MIT Press, 1994.

Graedel, T. E., and B. R. Allenby. *Industrial Ecology*. New York: Prentice-Hall, 1995.

Grier, James W. *Biology of Animal Behavior*. Toronto: Times Mirror/Mosby, 1984.

"Grizzly Plan Called Seriously Flawed." *National Parks* 68 (March/April 1994): 13.

Groves, Donald G. *Ocean World Encyclopedia*. New York: McGraw-Hill, 1980.

Gutnik, Martin J., and Natalie Browne-Gutnik. *Great Barrier Reef*. Austin: Raintree Steck-Vaughn Publishers, 1995.

Harms, Valerie, et. al. *Almanac of the Earth: The Ecology of Everyday Life*. National Audubon Society. New York: G. P. Putnam's Sons, 1994.

Harris, Jacqueline L. *Hereditary Diseases*. New York: Henry Holt, 1993.

Hazen, Robert M., and James Trefil. *Science Matters: Achieving Scientific Literacy.* New York: Anchor Books, 1991.

Heintzelman, Oliver H., and Richard M. Highsmith, Jr. *World Regional Geography.* 4th ed. Englewood Cliffs, NJ: Prentice-Hall, 1973.

Holmes, Gwendolyn, Ben Ramnarine Singh, and Louis Theodore. *Handbook of Environment and Technology.* New York: John Wiley & Sons, 1993.

Hunken, Jorie. *Ecology for All Ages: Discovering Nature Through Activities for Children and Adults.* Old Saybrook, CN: The Glove Pequot Press, 1994.

Huxley, Anthony. *Plant and Planet.* New York: The Viking Press, 1974.

Kamin, Michael A., and P. W. Rodgers, eds. *Dioxins in the Environment.* Washington, DC: Hemisphere Publishing, 1985.

Krebs, Charles J. *Ecology: The Experimental Analysis of Distribution and Abundance.* 4th ed. New York: HarperCollins, 1994.

———. *The Message of Ecology.* HarperCollins, 1988.

Kurzman, Dan. *Inside Union Carbide and the Bhopal Disaster: A Killing Wind.* New York: McGraw-Hill, 1987.

Ley, Willy. *The Poles.* New York: Time-Life Books, 1962.

Library of Pittsburgh, comp. *The Handy Science Answer Book.* Detroit: Visible Ink, 1994.

Livermore, Beth. "Just Where Have All the Frogs Gone?" *Smithsonian* (1992).

Lovelock, J. E. *Gaia: A New Look at Life on Earth.* Oxford: Oxford University Press, 1979.

Lutgens, Frederick K., and Edward J. Tarbuck. *The Atmosphere.* 2nd ed. Englewood Cliffs, NJ: Prentice-Hall, 1982.

Madsen, Axel. *Cousteau: An Unauthorized Biography.* New York: Beaufort Books, 1986.

Manahan, Stanley E. *Environmental Chemistry.* 4th ed. Chelsea, MI: Lewis Publishers, 1990.

Marro, Anthony, J., ed. *From Newsday Rush to Burn.* Washington, DC: Island Press, 1989.

Miller, Brian, Gerardo Ceballos, and Richard Reading. "The Prairie Dog and Biotic Diversity." *Conservation Biology* 8 (September 1994).

Miller, E. Willard. *Environmental Hazards: Toxic Waste and Hazardous Material.* Santa Barbara, CA: ABC-CLIO, 1991.

Miller, G. Tyler, Jr. *Living in the Environment.* 8th ed. Belmont, CA: Wadsworth Publishing, 1994.

Ming, L. "Fighting China's Sea of Sand." *International Wildlife* 18, no. 6 (1988): 38–45.

National Parks and Conservation Association. *Our Endangered Parks: What You Can Do to Protect Our National Heritage.* San Francisco: Foghorn Press, 1994.

Nebel, Bernard J., and Richard T. Wright. *Environmental Science: The Way the World Works.* 5th ed. Upper Saddle River, NJ: Prentice-Hall, 1996.

Norton, Bryan G. "The Yellowstone Complex." *Wilderness* 50 (Spring 1987): 26–30.

Noss, Reed F., and Allen Y. Cooperrider. *Saving Nature's Legacy: Protecting and Restoring Biodiversity.* Washington, DC: Island Press, 1994.

Ochoa, George, and Melinda Corey. *The Timeline Book of Science.* New York: Stonehenge Press, 1995.

Odum, Eugene. *Basic Ecology*. Philadelphia. Saunders College Publishing, 1983.

———. *Ecology and Our Endangered Life-Support Systems*. 2nd ed. Sunderland, MA: Sinauer Associates, 1993.

"Oil and Gas Drilling Threatens Grizzlies." *National Parks* 67 (May/June 1993): 13–14.

Orians, Gordon, and E.W. Pfeiffer. "Ecological Effects of the War in Vietnam." In *Global Ecology*, edited by Charles H. Southwick. Sunderland, MA: Sinauer Associates, 1985.

Parfit, Michael. "Diminishing Returns: Exploiting the Ocean's Bounty." *National Geographic Magazine* (November 1995).

Piller, Charles. *The Fail-Safe Society: Community Defiance and the End of American Technological Optimism*. New York: Basic Books, 1991.

Porter, K. R. *Herpetology*. Philadelphia: W. B. Saunders Company, 1972.

Postgate, John. *Microbes and Man*. 3rd ed. New York: Cambridge University Press, 1992.

Poten, Constance J. "A Shameful Harvest: America's Illegal Wildlife Trade." *National Geographic Magazine* (September 1991).

Pryor, Karen, and Kenneth S. Norris, eds. *Dolphin Societies: Discoveries and Puzzles*. Berkeley, CA: University of California Press, 1991.

Purves, W. K., and G. H. Orians. *Life: The Science of Biology*. Sunderland, MA: Sinauer Associates, 1983.

Rapport, D. J., C. L. Gaudet, and P. Calow. *Evaluating and Monitoring the Health of Large-scale Ecosystems*. Berlin: Springer-Verlag, 1995.

Raven, Peter. *Biology of Plants*. New York: Worth Publishers, 1981.

Ray, Dixy Lee, and Lou Guzzo. *Trashing the Planet*. New York: Harper Perennial, 1990.

Reading, R. P. "Toward an Endangered Species Reintroduction Paradigm: A Case Study of the Black-footed Ferret." Ph.D. diss., Yale University, 1993.

Reed, Alexander W. *Ocean Waste Disposal Practices*. Park Ridge, NJ: Noyes Data Corporation, 1975.

Revelle, Penelope, and Charles Revelle. *The Environment: Issues and Choices for Society*. 2nd ed. Boston: PWS publishers, 1984.

Ricklefs, Robert E. *The Economy of Nature*. 3rd ed. New York: W. H. Freeman, 1993.

Sackett, Russell. *Edge of the Sea*. Alexandria, VA: Time-Life Books, 1983.

Sale, Kirkpatrick. *The Green Revolution: The American Environmental Movement 1962–1992*. New York: Hill and Wang, 1993.

Satchell, Michael. "A New Battle Over Yellowstone Park." *U.S. News and World Report* 118 (March 1995): 34.

Schmidt-Nielsen, Knut. *How Animals Work*. Cambridge: Cambridge University Press, 1972.

Schwartz, Meryl. *The Environment and the Law*. New York: Chelsea House Publishers, 1993.

Silverstein, Alvin, and Virginia. *The Genetics Explosion*. New York: Four Winds, 1980.

Southwick, Charles H. *Ecology and the Quality of Our Environment*. 2nd ed. Boston: Prindle, Weber, and Schmidt, 1976.

Starr, Cecil, and Ralph Taggart. *Biology*. 6th ed. Belmont, CA: Wadsworth Publishing, 1992.

Steger, Will. *Saving the Earth. A Citizen's Guide to Environmental Action*. New York: Alfred A. Knopf, 1990.

Stephenson, Thomas A., and Anne Stephenson. *Life Between Tidemarks on Rocky Shores*. San Francisco: W. H. Freeman, 1972.

Stoddard, Charles. *Essentials of Forestry Practice*. New York: John Wiley and Sons, 1978.

Strahler, Arthur N., and Alan H. Strahler. *Elements of Physical Geography*. 3rd ed. New York: John Wiley and Sons, 1984.

Suzuki, David T., Anthony J. F. Griffiths, and Richard C. Lewontin. *An Introduction to Genetic Analysis*. 2nd ed. San Francisco: W. H. Freeman, 1981.

Tarbuck, Edward J., and Frederick K. Lutgens. *Earth Science*. 3rd ed. Columbus, Ohio: Charles E. Merrill, 1982.

Tate, Nicholas. *Sick Building Syndrome*. Far Hills, NJ: New Horizon, 1994.

Toynbee, Arnold. *The Industrial Revolution*. Boston: The Beacon Press, 1956.

Tyning, T. F. *A Guide to Amphibians and Reptiles*. Boston: Little, Brown and Co., 1990.

U.S. Congress, Office of Technology Assessment. *Facing America's Trash: What Next for Municipal Solid Waste*. OTA-O-424. Washington, DC: U.S. Government Printing Office, October, 1989.

U.S. Environmental Protection Agency. "In Projuctive Harmony—Environmental Impact Statements Broaden the Nation's Perspectives." Washington, DC: U.S. Government Printing Office, EPA-335.

Villee, Claude A., Eldra Pearl Solomon, and P. William Davis. *Biology*. Philadelphia: Saunders College Publishing, 1985.

Waldbott, George L. *Health Effects of Environmental Pollutants*. St. Louis, MO: C. V. Mosby, 1973.

Waldron, H. A., ed. *Metals in the Environment*. New York: Academic Press, 1980.

Wallace, Bruce. *The Search for the Gene*. Ithaca, NY: Cornell University Press, 1992.

Westman, Walter E. *Ecology, Impact Assessment, and Environmental Planning*. New York: John Wiley and Sons, 1985.

Wetzel, Robert G. *Limnology*. Philadelphia: Saunders College Publishing, 1975.

Wexo, John B. "Giant Pandas." *Zoobooks,* vol. 2, no. 11. San Diego, CA: Wildlife Education, Ltd., 1986.

Wiens, Herold J. *Atoll Environment and Ecology*. New Haven: Yale University Press, 1962.

Williams, G. C. "Gaia, Nature Worship and Biocentric Fallacies." *The Quarterly Review of Biology* vol. 67, no. 4 (December 1992).

Wilson, Edward O. *The Diversity of Life*. New York: W. W. Norton, 1992.

Wilson, Roberta, and James Q. Wilson. *Watching Fishes: Life and Behavior on Coral Reefs*. New York: Harper and Row, 1985.

Zug, George R. *Herpetology: An Introductory Biology of Amphibians and Reptiles*. San Diego: Academic Press.

# Photo Credits

b=bottom, t=top, c=center, l=left, r=right

## VOLUME I

**AP/Wide World Photos**: 33, 40(b), 54, 92; Michael Lipchitz/AP/Wide World Photos: 130; Tannen Maury/AP/Wide World Photos: 42(tl); Carlos Osorio/AP/Wide World Photos: 92(t); **Christine L. Coscioni**: 85, 120; Michael Durham/**Ellis Nature Photography**: 28(tr), 29, 87(c, r); Gerry Ellis/Ellis Nature Photography: 3(tl, tr, c, b), 7(t,b), 17, 28(tl); Roger Steene/Ellis Nature Photography: 127(t); **Garry D. Michael Resources**: 34(t); **Valerie Henschel**: 86; Vhldes/ **Lehtlkovh**: 93(r); **NASA**: 44(tl), 127(c); **NOAA**: 101(r); **John R. Patton**: 108; Dick George/**Phoenix Zoo Photo**: 82; A. Ramey/**PhotoEdit**: 113; **D. Robison**: 15(c); Anthony Mercieca/**Root Resources**: 64(tl); **UPI/Corbis— Bettman**: 89; **U.S. Department of Energy**: 25(t), 47(t); Robert Frerck/U.S. Department of Energy: 25(b); **U.S. Fish & Wildlife Service**: 78; Dave Clendenen/U.S. Fish & Wildlife Service: 79; **Visuals Unlimited**: 50; M. Abbey/Visuals Unlimited: 49(br); T. E. Adams/Visuals Unlimited: 49(t); J. Alcock/Visuals Unlimited: 125(tl); Patricia Armstrong/Visuals Unlimited: 107, 117(t), 122(t); Joel Arrington/Visuals Unlimited: 52; W. A. Banaszewski/Visuals Unlimited: 116, 123; Hal Beral/Visuals Unlimited: 62(t); Scott Berner/Visuals Unlimited: 23(t); Paul Bierman/Visuals Unlimited: 95; Bayard H. Brattstrom/Visuals Unlimited: 103(t); C. Brierley/Visuals Unlimited: 126(b); Veronika Burmeister/Visuals Unlimited: 22(c); D. Cavagnaro/Visuals Unlimited: 14(b), 56, 110; Kevin & Betty Collins/Visuals Unlimited: 76; A. J. Copley/Visuals Unlimited: 18(c), 103(b); John D. Cunningham/Visuals Unlimited: 4, 14(tr), 23(c), 49(bl), 67(t), 96, 104(tr), 105(tl), 124(tl, tc), 125(tc); Tom Edwards/Visuals Unlimited: 112(r); Stan W. Elems/Visuals Unlimited: 26; David H. Ellis/Visuals Unlimited: 55; Tony Freeman/Visuals Unlimited: 70(b); Barbara Gerlach/Visuals Unlimited: 6, 73(r); Mark E. Gibson/Visuals Unlimited: 102(b), 114, 117(c); Daniel W. Gotshall/Visuals Unlimited: 75; George Herben/Visuals

Unlimited: 30(c), 46(tc); Arthur R. Hill/Visuals Unlimited: (15(b), 42(tr), 105(br), 112(l); Historic/Visuals Unlimited: 45; Bill Kamin/Visuals Unlimited: 72; R. Kessel-G. Shih/Visuals Unlimited: 47(b); Kirtley-Perkins/Visuals Unlimited: 101(l); Frank Lambrecht/Visuals Unlimited: 15(tl); Stephen J. Lang/Visuals Unlimited: 81; Hank Levine/Visuals Unlimited: 44(tr), 103(r); LINK/Visuals Unlimited: 13(t), 30(t), 65, 98(r), 117(b); Ken Lucas/Visuals Unlimited: 83 (t, b); S. Maslowski/Visuals Unlimited: 1, 73(tr, bl); Steve McCutcheon/Visuals Unlimited: 20(c), 39, 73(tl), 105(tr); Joe McDonald/Visuals Unlimited: 5, 62(c), 68(c), 69, 124(b), 125(r, inset); D. Newman/Visuals Unlimited: 46(tr); Glenn M. Oliver/Visuals Unlimited: 64(tr); David M. Phillips/ Visuals Unlimited: 22(t), 49(c), 80; Martha Powell/Visuals Unlimited: 22(cr); Dick Poe/Visuals Unlimited: 67(b); Gary R. Robinson/Visuals Unlimited: 23(b); Len Rue, Jr./Visuals Unlimited: 119; Leonard Lee Rue III/Visuals Unlimited: 20(b); Kjell B. Sandved/Visuals Unlimited: 13(c); Science/ Visuals Unlimited: 12, 14(tl), 18(t), 53, 70(tl), 105(bl), 106, 112(b); John Serrao/Visuals Unlimited: 68(b); SIU/Visuals Unlimited: 31, 87(l), 97; John Sohlden/Visuals Unlimited: 57, 102(t), 104(b); Doug Sokell/Visuals Unlimited: 27, 70(tr), 98(tl), 104(tl), 126(t); Ron Spomer/Visuals Unlimited: 125(bl); Will Troyer/Visuals Unlimited: 40(t); Richard Walters/Visuals Unlimited: 62(b); W. J. Weber/Visuals Unlimited: 111, 125(bc); M. K. Wicksten/Visuals Unlimited: 109; Sylvan Wittwer/Visuals Unlimited: 34(b), 131(t, b); Chuck Nacke/**Woodfin Camp & Associates**: 93(c).

## VOLUME 2

**AP/Wide World Photos**: 12, 19, 23, 24(b), 28(r), 36(t), 40, 61, 75, 110(b), 129; **Gary Braasch**: 28(l); **Nancy Brown**: 35; **Christine L. Coscioni**: 124; Larry Foster/**EarthViews**: 32(b); Robert L. Pitman/EarthViews: 32(t), 71; Gerry Ellis/ **Ellis Nature Photography:** 15, 48, 76, 121(r); Jeremy

Stafford-Deitsch/Ellis Nature Photography: 105; **Greenpeace/ Estcourt**: 60(bl); **George Hall**: 133(r); **Emily Johnson**: 136, 137; Stuart Nicole/**Katz Pictures**: 99; **Maull and Fox**: 3; Anna E. Zuckerman/**PhotoEdit**: 55; **Planned Parenthood of NYC, Inc.**: 98(tc, tr); **Root Resources**: 74; John Kohout/Root Resources: 135; **Joe Thompson**: 34; **U.S. Department of Agriculture**: 109; **U.S. Department of Energy**: 51; Steve Hillebrand/**U.S. Fish & Wildlife Service**: 107(r); John & Karen Hollingsworth/U.S. Fish & Wildlife Service: 107(l); Walt Anderson/**Visuals Unlimited**: 13; Hank Andrews/Visuals Unlimited: 92(t); Patricia Armstrong/Visuals Unlimited: 104(c); Rudolph Arndt/ Visuals Unlimited: 25(t); Joel Arrington/Visuals Unlimited: 82(t), 83; R. Ashley/Visuals Unlimited: 10(t); Frank Awbrey/ Visuals Unlimited: 80(r); Bill Beatty/Visuals Unlimited: 6, 9(tl, tr, bl, br), 115(bc); Adrian Carton/Visuals Unlimited: 53(t); D. Cavagnaro/Visuals Unlimited: 10(b); Les Christman/Visuals Unlimited: 9l, 115(bl); A. J. Copley/ Visuals Unlimited: 19(c), 50; John D. Cunningham/Visuals Unlimited: 2(b), 11(c), 25(b), 92(b), 102(c, r), 111; Beth Davidow/Visuals Unlimited: 87; Brenda Dillman/Visuals Unlimited: 112; Derrick Ditchburn/Visuals Unlimited: 104(l); Stan W. Elems/Visuals Unlimited: 53(b); Don Fawcett/Visuals Unlimited: 95; John Flannery/Visuals Unlimited: 38; Dave Fleetham/Visuals Unlimited: 106(c); Brian Foss/Visuals Unlimited: 89; Tony Freeman/Visuals Unlimited: 49; Mark E. Gibson/Visuals Unlimited: 69(b), 88; Tim Hauf/Visuals Unlimited: 16(t); C. P. Hickman/Visuals Unlimited: 4, 73; Arthur R. Hill/Visuals Unlimited: 16(b), 86; J. A. Jackson/Visuals Unlimited: 101(t); Kirtley-Perkins/ Visuals Unlimited: 125; Stephen J. Lang/Visuals Unlimited: 104(r); R. Lindholm/Visuals Unlimited: 58(t); LINK/Visuals Unlimited: 2(t), 72; D. Long/Visuals Unlimited: 67(l); John Lough/Visuals Unlimited: 96, 97(t, b); S. Maslowski/Visuals Unlimited: 60(t), 106(l); Jane McAlonan/Visuals Unlimited: 66; James McCullagh/Visuals Unlimited: 80(l); Steve McCutcheon/Visuals Unlimited: 36(b), 78(t), 101(b); Joe McDonald/Visuals Unlimited: 58(b), 110(t), 127; Martin G. Miller/Visuals Unlimited: 82(b); Glenn M. Oliver/Visuals Unlimited: 69(t), 85; H. Oscar/Visuals Unlimited: 60(br); David M. Phillips/Visuals Unlimited: 11(t); Roger A. Powell/

Visuals Unlimited: 121(l); G. Prance/Visuals Unlimited: 14; S. W. Ross/Visuals Unlimited: 108; Leonard Lee Rue III/Visuals Unlimited: 43(t); Kjell B. Sandved/ Visuals Unlimited: 56; Science/Visuals Unlimited: 122, 131; 133(l); Joe Shute/Visuals Unlimited: 62; John Sohlden/ Visuals Unlimited: 27; Doug Sokell/Visuals Unlimited: 24(t); Ron Spomer/Visuals Unlimited: 39; Richard Thom/Visuals Unlimited: 126; Dick Thomas/Visuals Unlimited: 5, 67(b); Tom Ulrich/Visuals Unlimited: 43(b), 67(tc); Ken Wagner/ Visuals Unlimited: 78(b); R. Wallace/Visuals Unlimited: 1; E. Webber/Visuals Unlimited: 68; W. J. Weber/Visuals Unlimited: 79, 115(t); Bernard Boutrit/**Woodfin Camp & Associates**: 19(t); Anthony Rollo/Woodfin Camp & Associates: 63; William Strode/Woodfin Camp & Associates: 19(r); Adam Woolfit/Woodfin Camp & Associates: 138.

# VOLUME 3

**D. Allen**: 43(b); **AP/Wide World Photos**: 6, 9, 113; **CARE**: 82; **Christine L. Coscioni**: 5(b), 119(bl); Gerry Ellis/**Ellis Nature Photography**: 5(t), 18, 28, 29, 39, 109(t), 48(t); **Valerie Henschel**: 49; Hosoya/**IAEA**: 90; **John Patton**: 116; Tony Freeman/**PhotoEdit**: 106; Felicia Martinez/Photo-Edit: 104(r); Michael Newman/PhotoEdit: 104(l); David Young-Wolff/PhotoEdit: 45(tr); **Root Resources**: 105; Kenneth W. Fink/Root Resources: 100; **Royal Botanic Garden**: 10; **Kevin Schafer**: 91; **Mark Turner**: 117(t); **University of Wisconsin**: 118; **U.S. Department of Agriculture**: 122(t); **U.S. Evironmental Protection Agency**: 123(br); **Vireo**: 34; Walt Anderson/**Visuals Unlimited**: 32(t), 42(b), 97, 122(b); Joel Arrington/Visuals Unlimited: 93; Frank Awbrey/Visuals Unlimited: 41, 42(t); W. A. Banaszewski/Visuals Unlimited: 43(t), 51; Hal Beral/Visuals Unlimited: 44; James Alan Brown/Visuals Unlimited: 63; D. Cavagnaro/Visuals Unlimited: 81, 95(t); A. J. Copley/Visuals Unlimited: 83(b); John D. Cunningham/Visuals Unlimited: 56(t), 68(b), 75, 85(r), 119(tl), 87; Tom Edwards/Visuals Unlimited: 64(b); Ron David Farris/Visuals Unlimited: 26; Don W. Fawcett/Visuals Unlimited: 32(b), 77(tl, tr); Barbara Gerlach/Visuals Unlimited: 74; Daniel W. Gotshall/Visuals Unlimited: 96(r),

# VOLUME 4

Ziminski/ Visuals Unlimited: 18, 39; **Pictorial History Research**: 33; Newsweek/**Woodfin Camp & Associates, Inc.**: 108; **Mike Yamashita**: 19.

## VOLUME 5

**Diane Ali**: 36(b); **AP/Wide World Photos**: 8, 10, 15(b), 82, 97; Wesley Wong/AP/Wide World Photos: 93; **Myrleen Cate**: 83; Gerald Allen/**Ellis Nature Photography**: 121(t); Gerry Ellis/Ellis Nature Photography: 38(b), 54; **HMS Rose Foundation**: 61(b); **Donald Leopold**: 81; **Ocean Arks International**: 100; Tony Freeman/**PhotoEdit**: 60; Jeff Greenberg/PhotoEdit: 42(t); Alan Oddie/PhotoEdit: 124, 125; A. Ramey/PhotoEdit: 130, 131; David Young-Wolff/Photo Edit: 118; **Pictorial History Research**: 79(b); **U.S. Department of Energy**: 49; **Visuals Unlimited**: 79(t); T. E. Adams/Visuals Unlimited: 14; David S. Addison/Visuals Unlimited: 26(b); Amax/Visuals Unlimited: 105; R. F. Ashley/Visuals Unlimited: 133(b); Bill Beatty/Visuals Unlimited: 62(b); Bayard Brattstrom/Visuals Unlimited: 65; Veronika Burmeister/Visuals Unlimited: 18; D. Cavagnaro/Visuals Unlimited: 12, 22(b), 37(t), 120(r); A. J. Copley/Visuals Unlimited: 4(tl), 50, 69; Albert Copley/Visuals Unlimited: 63; D. Clayton/Visuals Unlimited: 42(b); John D. Cunningham/Visuals Unlimited: 4(b), 20(t, b), 21, 37(b), 62(t), 70, 78, 89, 92(b), 137(l); 138; Fisher/Visuals Unlimited: (9t); John S. Flannery/Visuals Unlimited: 77; John Gerlach/Visuals Unlimited: 28, 90; Mark E. Gibson/Visuals Unlimited: 43(c), 58, 61(tr), 116; Daniel W. Gotshall/Visuals Unlimited: 85; Jeff Greenberg/Visuals Unlimited: 133; Frank M. Hanna/Visuals Unlimited: 68, 86, 92(t); Mack Henley/Visuals Unlimited: 43(t); Dale Jackson/Visuals Unlimited: 104(b); Bill Kamin/Visuals Unlimited: 6; Alex Kerstitch/Visuals Unlimited: 27(t), 67; Kirtley-Perkins/Visuals Unlimited: 39(b), 41(r), 104(t); LINK/Visuals Unlimited: 34; M. Long/Visuals Unlimited: 3; John Lough/Visuals Unlimited: 4(tr), 40(b); Ken Lucas/Visuals Unlimited: 61(tl); Karl Maslowski/Visuals Unlimited: 109; David Matherly/Visuals Unlimited: 55; Jane McAlonan/Visuals Unlimited: 29; James B. McCullagh/Visuals Unlimited: 132; Joe McDonald/Visuals Unlimited: 7, 38(t, c), 73(t), 121(b), 137(r); Martin G. Miller/Visuals Unlimited: 76; Arthur Morris/Visuals Unlimited: 115;

K. G. Murti/Visuals Unlimited: 15(t); Bob Newman/Visuals Unlimited: 9(b); D. Newman/Visuals Unlimited: 39(t); W. Ormerod/Visuals Unlimited: 22(t); William Palmer/Visuals Unlimited: 43(b); N. Pecnik/Visuals Unlimited: 26(t), 87(t); G. Perkins/Visuals Unlimited: 40(t); Science/Visuals Unlimited: 25, 41(l), 64, 117; Stephen Sharnoff/Visuals Unlimited: 16(b); John Sohlden/Visuals Unlimited: 44, 117, 118(t); Doug Sokell/Visuals Unlimited: 27(b); Ron Spomer/Visuals Unlimited: 2, 36(t); David Stone/Visuals Unlimited: 1(t); Richard Thom/Visuals Unlimited: 72; Tom Ulrich/Visuals Unlimited: 56; Ken Wagner/Visuals Unlimited: 57; W. J. Weber/Visuals Unlimited: 13, 73(b), 87(b), 120(c); **Woodfin Camp & Associates**: 113; Bill Pierce/Woodfin Camp & Associates: 51; Roger Werth/Woodfin Camp & Associates: 128.

## VOLUME 6

**AP/Wide World Photos**: 7(t), 62; **Christine L. Coscioni**: 28(b); Gerry Ellis/**Ellis Nature Photography**: 31(t), 33(t, b), 60; Jeremy Stafford-Dietsche/Ellis Nature Photography: 1, 31(b), 32; **Harvard University**: 70; A. Ramey/**PhotoEdit**: 17; **Kevin Schafer**: 57(l, r); **Tom Stack**: 58; **U.S. Department of Energy**: 71; **Visuals Unlimited**: 39; Larry Blank/Visuals Unlimited: 68; Les Christman/Visuals Unlimited: 5; John D. Cunningham/Visuals Unlimited: 4, 73, 74(b); Beth Davidow/Visuals Unlimited: 2; Mark E. Gibson/Visuals Unlimited: 77(t); Jeff Greenberg/Visuals Unlimited: 48; Frank M. Hanna/Visuals Unlimited: 52; Tim Hauf/Visuals Unlimited: 7(b); William Jorgenson/Visuals Unlimited: 78; E. A. Kuttpapan/Visuals Unlimited: 13; Karl Maslowski/Visuals Unlimited: 66; Joe McDonald/Visuals Unlimited: 5(c), 67, 77(b); Leonard Lee Rue III/Visuals Unlimited: 64; Kjell B. Sandved/Visuals Unlimited: 46; Science/Visuals Unlimited: 12, 14, 15, 24, 29, 37; John Sohlden/Visuals Unlimited: 6; Richard Thom/Visuals Unlimited: 59; Jeanette Thomas/Visuals Unlimited: 23; **The White House**: 28(t); Gary Braasch/**Woodfin Camp & Associates**: 3, 19; Bernnard Boutrit/Woodfin Camp & Associates: 74(t); Roger Werth/Woodfin Camp & Associates: 35; James Wilson/Woodfin Camp & Associates: 75.

Title page illustration by Tom Cardamone for BBI.

# Subject Index

## BIOLOGY

Aerobic
Algae
Anaerobic
Bacteria
Evapotranspiration
Fern
Flowering Plant
Fungi
Gypsy Moth
Hybridization
Lichen
Mangroves
Migration
Photosynthesis
Phytoplankton
Plankton
Plant Pathology
Plants
Pollination
Respiration
Virus
Xenobiotic
Zooplankton

## CHEMISTRY

Aluminum
Carbon
Carbon Dioxide
Carbon Monoxide
Copper
Lead
Minerals
Oxygen
Ozone
pH
Plutonium
Uranium

## CLIMATE AND WEATHER

Atmosphere
Climate
Climate Change
Clouds
El Niño
Global Warming

Greenhouse Effect
Greenhouse Gas
Mesosphere
Meteorology
Ocean Current
Ozone Hole
Ozone Layer
Precipitation
Stratosphere
Troposphere
Weather

## ECOLOGICAL CONCEPTS

### General

Abiotic Factors
Biological Community
Bioregion
Biosphere
Commensalism
Detritus
Ecology
Ecosystem
Environment
Exotic Species
Extinction
Habitat
Hydrosphere
Keystone Species
Lithosphere
Mutualism
Native Species
Niche
Symbiosis

### Biogeochemical Cycles

Biogeochemical Cycle
Carbon Cycle
Nitrogen Cycle
Nitrogen Fixing
Oxygen Cycle
Water Cycle

### Biomes

Antarctica
Arctic
Barrier Islands
Biome

Coniferous Forest
Coral Reef
Deciduous Forest
Desert
Estuary
Forest
Galápagos Islands
Grasslands
Great Lakes
Intertidal Zone
Ocean
Pampas
Prairie
Rain Forest
Salt Marsh
Savanna
Taiga
Tropics
Tundra
Wetlands

### Ecological Succession

Climax Community
Pioneer Species
Succession

### Energy Transfer in Ecosystems

Autotroph
Biomass
Carnivore
Consumer
Decomposer
Decomposition
Energy Pyramid
Food Chain
Food Web
Herbivore
Omnivore
Parasitism
Predator
Producer
Trophic Level

## ENERGY

### General

Energy Efficiency
Fuel

**Sources of Energy**
Alternate Energy Sources
Breeder Reactor
Cogeneration
Fuel Wood
Gasohol
Geothermal Energy
Hydroelectric Power
Nuclear Fission
Nuclear Fusion
Nuclear Power
Ocean Thermal Energy
  Conversion (OTEC)
Photovoltaic Cell
Solar Energy
Solar Heating
Synthetic Fuel
Tidal Energy
Wind Power
**Fossil Fuels**
Alaska Pipeline
Automobile
Fossil Fuels
Natural Gas
Oil Drilling
Oil Shale
OPEC
Peat
Petrochemical
Petroleum

## ENVIRONMENTAL AGENCIES— INTERNATIONAL
International Atomic Energy
  Agency (IAEA)
International Union for the
  Conservation of Nature and
  Natural Resources (IUCN)
International Whaling
  Commission (IWC)
United Nations Commission on
  Sustainable Development
United Nations Conference on
  the Human Environment
United Nations Environmental
  Programme (UNEP)
World Bank

## ENVIRONMENTAL AGENCIES— UNITED STATES
Army Corps of Engineers

Atomic Energy Commission
  (AEC)
Bonneville Power Administration
Bureau of Land Management
Bureau of Reclamation
Council on Environmental
  Quality
Department of Agriculture
  (USDA)
Department of the Interior
Environmental Protection
  Agency (EPA)
Federal Energy Regulatory
  Commission (FERC)
Fish and Wildlife Service
Forest Service
National Oceanographic and
  Atmospheric Administration
  (NOAA)
National Park Service
National Weather Service
Office of Surface Mining,
  Reclamation, and Enforcement
Tennessee Valley Authority
  (TVA)

## ENVIRONMENTAL AGREEMENTS— INTERNATIONAL
Antarctic Treaty
Convention on International
  Trade in Endangered Species
  of Fauna and Flora (CITES)
Earth Summit
International Convention for the
  Regulation of Whaling (ICRW)
Law of the Sea Convention
Migratory Bird Treaty
Montreal Protocol

## ENVIRONMENTAL DISASTERS
Bhopal Incident
Chernobyl Accident
Dust Bowl
*Exxon Valdez*
Love Canal
Natural Disasters
Nuclear Weapons
Nuclear Winter
Radioactive Fallout
Three Mile Island

## ENVIRONMENTAL EDUCATION
Environmental Education
Natural History Museum

## ENVIRONMENTAL HEALTH
Bioaccumulation
Cancer
Carcinogen
Health and Disease
Health and Nutrition
Heavy Metals Poisoning
Hypoxia
Minamata Disease
Pathogen
Radiation Exposure

## ENVIRONMENTAL LAWS— UNITED STATES
Air Pollution Control Act (1955)
Clean Air Act
Clean Water Act
Comprehensive Environmental
  Response, Compensation, and
  Liability Act (CERCLA)
Container Deposit Legislation
Endangered Species Act
Federal Insecticide, Fungicide,
  and Rodenticide Act (FIFRA)
Hazardous Materials
  Transportation Act
Hazardous Substances Act
Marine Mammal Protection Act
Marine Protection, Research and
  Sanctuaries Act
Mineral Leasing Act
Mining Law of 1872
National Environmental Policy
  Act (NEPA)
National Pollutant Discharge
  Elimination System (NPDES)
Reclamation Act of 1902
Resource Conservation and
  Recovery Act (RCRA)
Safe Drinking Water Act
Solid Waste Disposal Act
Superfund
Surface Mining Control and
  Recovery Act (SMCRA)
Taylor Grazing Act

Toxic Substances Control Act
Wetlands Protection Act
Wild and Scenic Rivers Act
Wilderness Act

## ENVIRONMENTAL MOVEMENTS
Deep Ecology
Earth Day
Ecofeminism
Ecological Economics
Ecotourism
Environmental Justice
Gaia Hypothesis
Green Politics

## ENVIRONMENTAL POLICY
Cost Benefit Analysis
Environmental Impact Statement
Law, Environmental
Risk Analysis
Risk Assessment

## ETHICS AND VALUES
Biophilia
Conservation
Environmental Ethics
Frontier Ethic
NIMBY
Tragedy of the Commons

## EVOLUTION
Adaptation
Adaptive Radiation
Coevolution
Competition
Convergent Evolution
Evolution
Extinction
Mass Extinction
Natural Selection

## FOOD
### Agriculture
Agricultural Revolution
Agroecology
Aquaculture
Contour Farming
Crop Rotation
Green Revolution
Humus

Legumes
Livestock
Monoculture
No-Till Agriculture
Organic Farming
Subsistence Agriculture
Sustainable Agriculture
Terracing
### Fishing
Fishing, Commercial
Fishing, Recreational
Gill Net
### Food Shortages
Famine
Malnutrition
### Soil
Desertification
Erosion
Loam
Salinization
Soil
Soil Conservation
Topsoil

## GENETICS
DNA
Gene
Gene Bank
Gene Pool
Genetic Diversity
Genetic Engineering
Genetics

## GEOLOGY
Coal
Continental Drift
Glaciation
Glacier
Gullying
Ice Age
Plate Tectonics
Sediment
Sedimentation
Weathering
Volcanism

## INDIGENOUS PEOPLE
Hunter-Gatherer Society
Indigenous People

## LAND USE
### General
Conservation Easement
Debt for Nature Swap
Land Stewardship
Land Use
Multiple Use
Public Land
Restoration Biology
Sustainable Development
### Cities
Industrialization
Mass Transit
### Forests
Agroforestry
Clear-cutting
Defoliant
Deforestation
Fire Ecology
Forest Fire
Forest Products Industry
Forestry
Old-Growth Forest
Prescribed Burn
Silviculture
Softwood
Tree Farming
Understory
### Grazing
Grazing
Overgrazing
Rangeland
### Mining
Mining
Open-Pit Mining
Smelter
Strip Mining
Surface Mining
Tailings

## NATURAL RESOURCES (CONCEPT OF)
Natural Resources
Nonrenewable Resources
Renewable Resources

## PERSONS IMPORTANT IN ENVIRONMENTALISM
Audubon, John James
Carson, Rachel Louise
Cousteau, Jacques-Yves

Darwin, Charles Robert
Douglas, Marjory Stoneman
Dubos, Rene
Fossey, Dian
Hardin, Garrett
Leopold, Aldo
Lovelock, James
MacArthur, Robert H.
Mead, Sylvia Earle
Mexia, Ynes
Morgan, Ann
Muir, John
Odum, Eugene Pleasants
Patrick, Ruth
Pinchot, Gifford
Roosevelt, Franklin Delano
Roosevelt, Theodore
Telkes, Maria
Thoreau, Henry David
Wilson, Edward Osborne

## PEST CONTROL
Biological Control
DDT
Fungicide
Herbicide
Insecticide
Integrated Pest Management
Pest Control
Pesticide
Rodenticide

## PHYSICS
Electricity
Electromagnetic Field
Electromagnetic Spectrum
Radiation
Radioactivity
Thermodynamics, Laws of
Ultraviolet Radiation
X rays

## POLLUTION
### General
Agricultural Pollution
Best Available Control
  Technology (BACT)
Dioxin
*Global 2000 Report, The*
Mercury
PCBs

Pollution
Pollution Permits
Primary Pollution
### Air Pollution
Acid Rain
Aerosol
Air Pollution
Asbestos
Catalytic Converter
CFCs
Freon
Hydrocarbon
Labeling, Environment
Life-Cycle Assessment
Methane
Nitrogen Oxide
Oil Pollution
Ozone Pollution
Particulates
Radon
Scrubber
Sick Building Syndrome
Smog
Sulfur Dioxide
### Noise Pollution
Noise Pollution
### Water Pollution
Algal Bloom
Biochemical Oxygen Demand
Bioremediation
Detergent
Dissolved Oxygen
Dredging
Effluent
Eutrophication
Leaching
Marine Pollution
Nonpoint Source
Oil Spills
Phosphate
Point Source
Runoff
Saltwater Intrusion
Thermal Water Pollution
Water Pollution
Water Quality Standards

## POPULATION GROWTH
Age Structure
Carrying Capacity
Demography

Exponential Growth
Family Planning
Fertility Rate
Infant Mortality
Limits to Growth
Overpopulation
Population Growth
Zero Population Growth

## WASTE MANAGEMENT
### General
Biodegradable
Ocean Dumping
Recycling, Reducing, Reusing
Seabed Disposal
Source Reduction
Waste Management
Waste Reduction
### Hazardous Waste
Fly Ash
Hazardous Waste
Hazardous Waste Management
Hazardous Wastes, Storage and
  Transportation of
Industrial Waste Treatment
Medical Waste
Radioactive Waste
Toxic Waste, International Trade in
Toxic Waste
### Solid Waste
Composting
Garbage
Landfill
Photodegradable Plastics
Plastic
Solid Waste
Solid Waste Incineration
### Wastewater
Septic Tank
Sewage
Sewage Treatment Plant
Sludge
Wastewater
Wastewater, Primary, Secondary,
  and Tertiary Treatment of
Wastewater Treatment Plant

## WATER RESOURCES
Aquifer
Artesian Well
Chlorination

Dams
Desalination
Floodplain
Hydrology
Infiltration
Irrigation
Reservoir
Riparian Land
Riparian Rights
River Basin
Salinity
Surface Water
Water, Drinking
Water Rights
Watershed
Water Table
Water Treatment
Zone of Saturation

## WILDLIFE AND ENDANGERED SPECIES

Amphibian
Animal Rights
Bald Eagle
Biodiversity

Birds
California Condor
Captive Propagation
Dolphins/Porpoises
Elephants
Endangered Species
Fish
Fish Ladder
Giant Panda
Gorilla
Grizzly Bear
Habitat Loss
Hunting
Insect
Invertebrate
Ivory-Billed Woodpecker
Mammals
Passenger Pigeon
Pet Trade
Poaching
Reptile
Restoration Biology
Salmon
Sea Turtle
Seals and Sea Lions
Species

Species Diversity
Spotted Owl
Subspecies
Tiger
Tuna
Vertebrate
Whales
Wildlife
Wildlife Conservation
Wildlife Management
Wildlife Rehabilitation

## WILDLIFE REFUGES, PARKS, AND ZOOS

Arctic National Wildlife Refuge
Everglades National Park
National Forest
National Grassland
National Parks
National Seashore
National Wildlife Refuge
Yellowstone National Park
Yosemite National Park
Zoo

# General Index

## ▶ A

abiotic factors, **1**:2-3
  phosphorus as, **5**:9
abyssal plains, **4**:83
acacia trees, **4**:36-37
acid precipitation, **1**:4, 104
  caused by burning coal, **1**:107
acid rain, **1**:4-6, 4*c*, 16, 104; **5**:106;
    **6**:44, 54
  aluminum released by, **1**:27, 29
  caused by burning coal, **1**:107
  Clean Air Act and, **1**:19
  nitrogen dioxide contributing to, **4**:69
  sulfur dioxide and, **5**:129
active solar heating, **5**:112, 114
acute diseases, **3**:55
adaptations, **1**:6-8; **2**:41-42
  in biological communities, **1**:61
  coevolution and, **1**:109
  convergent evolution and, **1**:125
  of fish, **2**:105-6
  of herbivores, **3**:64
adaptive radiation, **1**:8
additives, **1**:86*l*, 87
advanced sewage treatment, **5**:102
aeration, **3**:96*l*, 97
aerobic organisms, **1**:8
aerosols, **1**:8-9; **4**:111
aesthetics, **3**:105*l*
Agenda 21, **6**:27
Agent Orange, **2**:12-13, 28; **3**:63
age-specific fertility rate, **2**:103
age structure, **1**:9-10, 9*g*
  fertility rates and, **2**:103
agricultural pollution, **1**:10-13
agricultural products, **3**:34*l*
agricultural revolution, **1**:13; **3**:79
  population growth and, **5**:34
agriculture
  agroecology and, **1**:14-15
  agroforestry and, **1**:15-16
  contour farming, **1**:122
  crop rotation in, **1**:131
  deforestation and, **2**:13-14
  Dust Bowl and, **2**:37-38
  erosion of topsoil and, **2**:80-81
  Everglades damaged by, **2**:88
  gene banks for, **3**:10-11

  genetic diversity and, **1**:58
  Green Revolution in, **3**:38-39
  habitat destruction and, **5**:34
  insects and, **3**:86-87
  irrigation used in, **3**:97-100
    Reclamation Act and, **5**:59
    salinization and, **5**:84
  monoculture in, **1**:97; **4**:31-32; **5**:2
  no-till, **4**:74-75
  organic farming, **4**:99-101
  pest control in, **4**:121-24
    herbicides used in, **3**:62
  pollution resulting from, **1**:10-13
  on prairies, **5**:36
  silviculture and, **5**:104-5
  soil conservation for, **5**:110-11
  source reduction of solid wastes
    from, **5**:117-18
  subsistence, **5**:125-26
  sustainable, **5**:135
  terracing in, **6**:7-8
  tree farming, **6**:18-19
Agriculture, U.S. Department of (USDA),
  **2**:17-19
  food pyramid of, **3**:60, 60*c*
  Forest Service of, **2**:126-28; **4**:39-40;
    **5**:15-16, 44
  public lands and, **4**:36
  Soil Conservation Service of, **5**:110
  soils classified by, **5**:109
  sustainable agriculture program of,
    **4**:100
agroecology, **1**:14-15
agroforestry, **1**:15-16
air, *see* atmosphere
air pollution, **1**:16-19; **5**:30
  caused by fossil fuel use, **1**:24
    coal burning, **1**:107
  control of
    Best Available Control Technology
      in, **1**:53
    catalytic converters for, **1**:90
    scrubbers for, **5**:91; **6**:38
  from incinerators, **5**:117
  legislation on, **1**:20, 94-96
    *see also* Clean Air Act
  pollutants, **1**:17*t*, 19*c*
    fly ash, **2**:116
    from fuel wood, **2**:134
    nitrogen dioxide, **4**:68-69

    ozone, **4**:108-9, 112-13
    particulates, **4**:116-17
    sulfur dioxide, **5**:129
    primary, **5**:41*t*
  sick building syndrome and, **5**:102-3
  smog, **5**:106
  in troposphere, **6**:22
Air Pollution Control Act (U.S., 1955),
  **1**:20
air (barometric) pressure, **1**:43-44; **6**:53
Alaska
  Arctic National Wildlife Refuge in,
    **1**:40-41; **4**:54
  bald eagles in, **1**:50
  oil spill in (1989), **2**:96; **4**:95
  wilderness in, **6**:62
Alaska National Interest Conservation
    Lands Act (U.S., 1980), **4**:54
Alaska Pipeline, **1**:20-21
  oil spill and (1989), **2**:96
alcohol
  in gasohol, **3**:7
  produced by anaerobic organisms,
    **1**:30
algae, **1**:21-23
  algal blooms of, **1**:24
  as autotrophs, **1**:47
  blue-green, **1**:48; **5**:12, 119
  *Emiliania*; **3**:36
  eutrophication and, **6**:46
  kelp, **3**:102*l*
  in lichens, **2**:43; **3**:118
  Patrick's study of, **4**:118-19
  phytoplankton, **5**:14
  plankton, **5**:17
  as plants, **5**:19-20
algal blooms, **1**:11, 24; **2**:85; **5**:15, 100;
    **6**:46
alginic acid, **1**:23
alien species, *see* exotic species
allergies, **5**:102
alligators, **4**:62, 64; **6**:67
allopatric speciation, **5**:121
alloys, **3**:116*l*
alpine glaciers, **3**:21-23
alternative energy sources, **1**:24-27;
    **2**:50-52
  in conservation of fossil fuels,
    **1**:117-18
  global warming and, **3**:27

ocean thermal energy conversion,
  4:89-90
solar energy as, 5:111-13
tidal energy, 6:11-12
wind power, 6:70-72
aluminum, 1:27
  recycling of, 5:60
  released by acid rain, 1:29
Alvarez, Walter, 4:15
American chestnut tree, 2:119
American Museum of Natural History
  (New York), 4:59
American Tree Farm System, 2:126
amino acids, 3:57
*Amoco Cadiz* (oil tanker), 4:95, 96
amphibians, 1:27-30; 6:31-32
anadromous species, 5:84
anaerobes, 1:30
anaerobic organisms, 1:30
ancient forests, *see* old-growth forests
anemia, 3:75
angiosperms, *see* flowering plants
anglers, 2:109
animal husbandry, 3:122
animal rights, 1:31
animals, 5:119
  amphibians, 1:27-30
  in Antarctica, 1:32
  in Arctic, 1:38-39
    Arctic National Wildlife Refuge for,
      1:40-41
  bacteria inside of, 1:50
  biomes and, 1:66
  birds, 1:72-74
  captive, 2:34, 35*l*
  captive propagation of, 1:81-82
  carnivores, 1:87
  domesticated, 6:64*l*
  ectotherms, 2:22*l*, 23
  endangered, 2:58-61
  in Everglades, 2:88
  evolution of, 2:4, 89-91
  fishes, 2:105-7
  in forests, 2:9-10, 119-22
  fungicides for diseases of, 2:139
  of Galápagos Islands, 3:3-4
  of grasslands, 3:30
  herbivores, 3:64
  hunting of, 3:67-68
    poaching, 5:26-27
  hybrids, 3:68
  insects, 3:84-88
  invertebrates, 3:95-97
  livestock, 3:122-23
  mammals, 4:4-6
  migrations of, 4:22-26
    spawning grounds and, 4:24*l*

omnivores, 4:98
  as pets, 5:6-8
  predators, 5:37-38
  preserves for, 1:32*l*
  in rain forests, 5:54-55
  reptiles, 5:65-67
  respiration by, 5:71
  in riparian lands, 5:73
  in salt marshes, 5:86
  in savannas, 5:90
  in taigas, 6:2-3
  in tundras, 6:25
  vertebrates, 6:30-33
  in wetlands, 6:55
  wildlife, 6:63-65
    conservation of, 6:65-68
    management of, 6:68-69
    rehabilitation of, 6:69-70
  in zoos, 6:77-78
  *see also* birds; endangered species;
    fishes; mammals
Animal Welfare Act (U.S., 1966), 1:31
annuals (plants), 2:22*l*, 23
Antarctica, 1:32-33
  hole in ozone layer over, 4:109-12
Antarctic Circumpolar Current, 4:86
Antarctic Treaty (1959), 1:33-34
anthracite coal, 1:106; 2:130
anthropology, 4:58, 59*l*
antibiotics, 2:36*l*, 37
ants, 4:36-37
aquaculture, 1:34-35
aquatic ecosystems, 1:11*l*
  algae in, 1:22
  estuaries, 2:81-85
  keystone species in, 3:102
  phosphorus in, 5:10
  phytoplankton in, 5:13-15
aquatic food webs, 2:44*c*
aquatic habitats, 3:44-46
aquifers, 1:35-38; 3:98; 5:134; 6:15
  artesian wells for water from, 1:42
  leachates in, 3:106
  runoff from, 5:77
  saltwater intrusion into, 5:88-89
  water tables of, 6:50-51
Arabian oryx, 1:81
arable soils, 6:13, 14*l*
Aransas National Wildlife Refuge
  (Texas), 4:53
arboreal (tree) habitats, 3:46
*Archaeopteryx* (prehistoric bird), 1:72
archipelagos, 3:3*l*
Arctic, 1:38-39
Arctic Circle, 1:38
arctic hares, 6:2

Arctic National Wildlife Refuge (ANWR;
  Alaska), 1:21, 40-41; 4:54
area mining, 5:124
Argentina, pampas of, 4:114-15
Aristotle, 2:20, 41
arithmetic growth, 5:35
  *see also* linear growth
Army Corps of Engineers (U.S.), 1:41-42;
  4:12
arsenic, 3:61
arson, 2:123
artesian wells, 1:42
Arthropoda, 3:96
  *see also* insects
asbestos, 1:42-43; 4:28; 5:103
ascomycetes (sac fungi), 2:137
asexual reproduction, 5:120-21
asphalt, 5:4
asthma, 4:113*l*
Aswan High Dam (Egypt), 1:129; 2:113;
  3:72
atmosphere, 1:43-45; 4:20*c*
  air pollution in, 1:16-19; 5:30
  carbon cycle in, 1:60
  carbon dioxide in, 1:84, 85*g*
    sources of, 3:25-26
  climate and, 1:99-100
  clouds in, 1:103-4
  composition of, 4:105*c*
  Gaia hypothesis on, 3:2
  greenhouse effect and, 3:34-36
  greenhouse gases in, 1:82-83;
    3:26-27, 27*c*, 36*t*
  impact of fossil fuels on, 1:101*g*
  mesosphere, 4:19-20
  meteorology and, 4:20-21
  National Oceanic and Atmospheric
    Administration and, 4:41-42
  nitrogen in, 4:67, 68
  nitrogen dioxide in, 4:69
  oxygen in, 4:106, 108
  ozone in, 4:108-9
  ozone layer in, 4:111-12
    hole in, 4:109-11
    ultraviolet radiation and, 6:27
  photosynthesis and, 5:12
  pollutants in
    acid rain, 1:4
    aerosols, 1:8-9
    CFCs, 1:90-91
    PCBs, 4:119-20
    primary, 5:41*t*
    radioactive fallout, 5:48
    volcanic ash, 5:41
  precipitation in, 5:37
  stratosphere, 5:123-24
  troposphere, 6:22-23

bisons (American buffalo), **2**:59; **4**:6
bituminous coal, **1**:106
black-footed ferrets, **3**:103
black light, **6**:26
Black Plague, **1**:50
blubber, **1**:32*l*; **6**:57
blue-gray gnatcatcher (bird), **4**:63
blue-green algae (cyanobacteria), **1**:48;
 **5**:12, 119
blue whales, **3**:90; **6**:57
bogs, **1**:38-39; **6**:24*l*
 peat bogs, **4**:121
Bolivia, **2**:8
Bonneville Power Administration (BPA),
 **1**:74-75
boreal forests, *see* taigas
Borlaug, Norman E., **3**:38, 39
botanists, **4**:22*l*
bottled water, **5**:82
Brady, Ray, **4**:66
brain dysfunction, **3**:116*l*
Brazil
 pampas of, **4**:114-15
 United Nations Earth Summit held in,
 **6**:27-28
breakwaters, **2**:81*l*
breathing, **5**:71
breeder reactors, **1**:75; **5**:25, 26
breeding, **4**:61*l*
bromeliads, **1**:2*l*
bronze, **5**:105
brown algae, **1**:23
brown pelicans, **6**:67
bryophytes (plants), **5**:20
Buffalo Creek (West Virginia), **5**:96
buildings
 energy efficiency of, **2**:64
 sick building syndrome, **5**:102-3
 solar heating of, **5**:112-14; **6**:6
Bureau of..., *see under main part of*
 *name*
burning, *see* fires; incineration
burrows, **3**:46*l*, 47
Bush, George, **4**:32
by-products, **1**:5*l*

◗ **C**

cadmium, **3**:61-62
caecilians, **1**:27, 28
Caesar, Julius (emperor, Rome), **3**:112
calcareous formations, **1**:126, 127*l*
Caldicott, Helen, **1**:123, 124
California
 earthquakes in, **4**:55

Mono Lake in, **5**:45
 San Andreas Fault in, **5**:25
 Sequoia National Park in, **4**:46
 wind power used in, **2**:52
 Yosemite National Park in, **4**:33-34,
 43; **5**:16; **6**:75
California condors, **1**:78-80; **2**:62-63;
 **6**:69, 78
California Desert Protection Act
 (U.S., 1994), **6**:62
calories, **3**:58*l*, 60
calving season, for whales, **6**:57, 58*l*
Campbell, H.W., **2**:37-38
Canada, Great Lakes in, **3**:33-34; **6**:47
cancer, **1**:80-81
 causes of
 carcinogens, **1**:86-87
 electromagnetic fields, **2**:54
 radiation, **5**:33, 47; **6**:11
 radon, **5**:52-53
 tobacco smoke, **5**:102-3
 trihalomethanes, **6**:48
 ultraviolet radiation, **4**:108
 treatments for, **1**:80*t*
canopy layer (in forests), **2**:9, 120; **5**:53
Cape Hatteras National Seashore
 (North Carolina), **4**:48
captive animals, **2**:34, 35*l*
captive propagation, **1**:81-82; **6**:69
 of California condors, **1**:79-80;
 **2**:62-63
 Fish and Wildlife Service programs in,
 **2**:107
carbohydrates, **1**:82; **3**:57-58; **5**:12*l*
 in photosynthesis, **5**:11-12
carbon, **1**:82-83
 in hydrocarbons, **3**:69
carbon black, **1**:82
carbon cycle, **1**:60, 83-84
carbon dating, **5**:51-52
carbon dioxide, **1**:18-19, 82-85
 in atmosphere, **1**:85*g*
 deforestation and, **2**:15-16
 released by burning of fossil fuels,
 **1**:101-2, 101*g*
 sources of, **3**:25-26
 in carbon cycle, **1**:60
 as greenhouse gas, **3**:35-37
 in photosynthesis, **5**:11-12
 produced by bacterial decomposers,
 **1**:49-50
carbonization, **2**:129*l*
carbon monoxide, **1**:18, 85-86
 catalytic converters for, **1**:90
 hypoxia and, **3**:75
carbon monoxide poisoning, **1**:85-86
carcinogens, **1**:80-81, 86-87

DNA damaged by, **2**:31
 in tobacco smoke, **5**:102-3
carnivores, **1**:87, 119, 120; **5**:84*l*; **6**:20-21
 amphibians as, **1**:29
 bioaccumulation of DDT in, **1**:51-52
 in food chains, **2**:117
 insects as, **3**:84-85
 seals and sea lions, **5**:93-95
 tigers, **6**:12-13
carrying capacity, **1**:88; **2**:93; **6**:69
 of human population, **2**:132
 limits to growth and, **3**:120
 overpopulation and, **4**:104
 population growth and, **5**:33-34
 of rangelands, **5**:58
Carson, Kit, **5**:4
Carson, Rachel Louise, **1**:xi, 88-90; **2**:6;
 **3**:37; **4**:65, 99
Carter, Jimmy, **5**:42, 130
cartilage, **2**:105*l*
cartilaginous fish, **2**:105*l*; **6**:31*l*
catalytic converters, **1**:47, 90
 to reduce acid rain, **1**:5
 to reduce air pollution, **1**:19
 to reduce carbon monoxide, **1**:86
 to reduce nitrogen dioxide, **4**:69
cattle ranching, **2**:14-15
*Cavtat* (Yugoslav ship), **1**:92-93
cells
 DNA in, **2**:29-32
 kingdoms based on structure of,
 **5**:118-19
 respiration and, **5**:70-71
cellular respiration, **5**:70, 71
cellulose, **5**:12, 138*l*
Central America, cholera in, **4**:118
central nervous systems, **6**:30
Central Valley Project (CVP; California),
 **1**:77
CERCLA, *see* Comprehensive
 Environmental Response,
 Compensation, and Liability Act
cesspools, **5**:100; **6**:40
CFCs (chlorofluorocarbons), **1**:9, 19,
 44-45, 90-91; **2**:131
 Clean Air Act on, **1**:96
 as greenhouse gases, **3**:27, 35-37
 Montreal Protocol on, **4**:32
 ozone layer and
 hole in, **4**:20, 106, 109, 111-12
 ultraviolet radiation and, **6**:27
chain reactions, **4**:76
channelization, **2**:113
channels, in estuaries, **2**:81
chaparrals, **2**:25, 26*l*
charcoal, **2**:134
Chase, Martha, **2**:30

coniferous trees, **5**:106*l*, 107; **6**:2
conifers, **1**:113
conservation, **1**:116-18; **5**:64
  environmental education and, **2**:67-68
  International Union for the
    Conservation of Nature and
    Natural Resources, **3**:91-92
  legislation on, **5**:69-70
  to reduce agricultural pollution, **1**:12
  of soil, **5**:110-11
   terracing for, **6**:7-8
  Tennessee Valley Authority actions
    on, **6**:7
  of water, **1**:77
  of wildlife, **6**:65-68
conservation easements, **1**:118
Conservation International
  (organization), **2**:7-8
conservation organizations, *see*
  environmental organizations
Consumer Products Safety Commission
  (U.S.), **3**:51
consumers (human)
  energy efficiency and, **2**:64
  environmental labeling for, **3**:104
  food labeling for, **3**:60
  USDA programs for, **2**:18
consumers (organisms), **1**:118-20, 119*t*;
  **2**:42, 116-17; **5**:63; **6**:20-21
  decomposers as, **2**:10
  in energy pyramids, **2**:65-66
  insects as, **3**:84-85
  producers and, **5**:41
container deposit legislation, **1**:120-21
continental drift, **1**:121-22
  plate tectonics and, **5**:23-25
continental glaciers (ice sheets), **3**:23
continental shelves, **3**:114*l*
contour farming, **1**:122; **2**:81; **5**:109, 111
contour mining, **5**:124
Control of Transboundary Movements of
  Hazardous Wastes and Their
  Disposal, United Nations treaty
  on (1989), **6**:17
Convention on International Trade in
  Endangered Species of Wild Fauna
  and Flora (CITES), **1**:122-24; **2**:61,
  96; **5**:8
  on elephants, **2**:56, 59
  on poaching, **5**:26-27
  on sea turtles, **5**:96
  on tigers, **6**:13
convergent evolution, **1**:124-25
convulsions, **3**:116*l*
coolants, **5**:31*l*
copper, **1**:126; **5**:105
  mining of

  open-pit, **4**:99; **5**:132
  tailings from, **6**:3
coral, **1**:21*l*, 22
coral reefs, **1**:126-28
  biodiversity in, **1**:58
  collecting, **5**:7
  Darwin on, **2**:3
core meltdowns, **6**:11*l*
Coriolis effect, **4**:86*l*
cormorants, **3**:3*l*
corrosion, **3**:115, 116*l*; **6**:4*l*
Cortés, Hernando, **6**:77
cost-benefit analysis, **1**:128-29; **3**:110;
  **5**:74
  Endangered Species Act and, **2**:62
cost-benefit ratios, **1**:128-29
cottony cushion scales (insects), **1**:64
Council on Environmental Quality (CEQ;
  U.S.), **1**:129-30; **4**:38
  environmental impact statements and,
    **2**:71, 72, 74
  Environmental Protection Agency
    and, **2**:77
  Global 2000 Report of, **3**:25
Council of State Governments (CSG),
  **6**:17
Cousteau, Jacques-Yves, **1**:130-31
Cousteau Society, **1**:131
Cretaceous mass extinction, **4**:14-15
Crick, Francis, **2**:30
crocodilians, **5**:65-66
crop rotation, **1**:12, 131; **4**:100-101; **5**:2;
  **6**:14
  to prevent erosion, **5**:109
  for soil conservation, **5**:110-11
  in sustainable agriculture, **5**:135
cross-pollination of plants, **5**:29
crude oil (petroleum), **2**:132; **5**:3
  refining of, **5**:5-6
  *see also* petroleum
crusts, **6**:44*l*
cultural control of pests, **4**:101, 124
cultural hazards, **3**:57
cultures (human)
  ethnobotanists and, **4**:22*l*
  of indigenous people, **3**:76-78
Curie, Marie, **5**:51
currents, *see* ocean currents
cyanobacteria, **1**:48; **5**:12
cystic fibrosis, **3**:14*l*

# ❱ D

dams, **2**:2-3; **3**:98
  Bureau of Reclamation for, **1**:76-77

endangered species and, **2**:62
fish ladders and, **2**:111-12; **4**:26; **5**:86
hydroelectric power from, **2**:50, 134;
  **3**:69-72
  Bonneville Power Administration
    and, **1**:74-75
  Tennessee Valley Authority and,
    **6**:6-7
  reservoirs formed by, **5**:67-68
  sediment behind, **5**:96
  water rights and, **5**:68-69
Darwin, Charles Robert, **1**:8, 111;
  **2**:3-5, 89
  on competition, **3**:79
  in Galápagos Islands, **3**:4
  on natural selection, **4**:61
DDT (dichlorodiphenyl trichloroethane),
  **1**:12; **2**:5-7; **3**:88; **4**:122-23; **5**:1
  bioaccumulation of, **1**:55
  Carson's warnings on, **1**:88-90
  impact of
    on bald eagles, **1**:51-52; **6**:67
    on California condors, **1**:79
deaths
  from diseases, **3**:55*c*
  from indoor air pollution, **5**:103
  infant mortality and, **3**:82-83
  in risk assessments, **5**:75-76
debris, **2**:120*l*
debt for nature swaps, **2**:7-8
decibel scale, **4**:70, 70*c*; **5**:30-31
deciduous forests, **1**:2; **2**:8-10, 120
  American chestnut tree in, **2**:119
  defoliants used on, **2**:12
  tropical and temperate, **2**:121
deciduous trees, **5**:106*l*, 107
decomposers (organisms), **1**:119-20;
  **2**:10-11, 28, 42; **5**:63
  bacteria as, **1**:49-50
  biodegradable materials broken
    down by, **1**:56
  in deciduous forests, **2**:8
  in food chains, **2**:116-17
  in forests, **2**:120
  fungi as, **2**:135, 136
  used in sewage treatment, **5**:101
decomposition, **2**:11
  of biodegradable materials, **1**:56
  in sewage treatment, **5**:101
  thermal water pollution and, **6**:8
deep ecology, **2**:11-12
deep-well injection, **5**:82, 83*l*; **6**:38
deer, **4**:115
defoliants, **2**:12-13; **3**:62
  *see also* herbicides
deforestation, **1**:60; **2**:13-16

euphorbia plants, **1**:66, 67*l*
eutrophication, **1**:11; **2**:85; **5**:10; **6**:46
evaporation, **2**:21*l*, 86; **3**:98; **6**:43
    in desalinization, **2**:20
evapotranspiration, **2**:86
Everglades National Park (Florida), **2**:35,
    86-88; **4**:44
    panthers in, **5**:38
    saltwater intrusion into, **5**:88
evolution, **2**:89-91
    adaptation in, **1**:6-8
    adaptive radiation in, **1**:8
    of bacteria, **1**:48
    of birds, **1**:72
    coevolution and, **1**:108-9
      mutualism and, **4**:37
      parasitism and, **4**:116
    convergent, **1**:124-25
    Darwin on, **2**:3-5
    extinctions and, **2**:94-96
      mass extinctions, **4**:13-15
    within gene pools, **3**:11
    genetic diversity and, **3**:12
    of humans, **1**:69
    of mammals, **4**:4-6
    natural selection in, **4**:61-62
      in resistance to pesticides, **4**:123
    speciation and, **5**:121-23
exotic pets, **5**:6-8
exotic species, **2**:59-60, 91-92; **6**:66
    gypsy moths as, **3**:42-43
    impact of food webs of, **2**:119
    in national grasslands, **4**:41
    native species versus, **4**:54
    overpopulation of, **4**:104
    predators as, **5**:2
exponential growth, **1**:88; **2**:93, 93*g*;
    **5**:34*g*
exports, **2**:14*l*
extended forecasts, **6**:54*l*
extinctions, **1**:xi, 59, 60; **2**:94-96, 95*g*;
    **5**:123; **6**:64, 65*l*, 66
    animals saved from, **6**:67
    of birds, **1**:74
      passenger pigeons, **4**:117-18
    of endangered species, **2**:58
    Global 2000 Report on, **3**:25
    habitat loss and, **3**:47
    impact of food webs of, **2**:119
    of mammals, **4**:6
    mass extinctions, **4**:13-15, 13*c*
    speciation and, **5**:119*c*
Exxon (firm), **2**:96
*Exxon Valdez* (oil tanker), **1**:21, 41; **2**:96;
    **4**:95

# ▶ F

fallout, radioactive, **5**:47-48
family planning, **2**:98-99
    contraception use and, **2**:98*g*
famines, **2**:71, 99-100; **4**:3-4
farm cooperatives, **2**:18*l*
farming, *see* agriculture
farmland, soil conservation in, **5**:110
fats, **3**:58
fauna, *see* animals
fecal coliform bacteria, **6**:47, 48*l*
Federal Energy Regulatory Commission
    (FERC; U.S.), **2**:100-101
federal government, *see* United States
Federal Insecticide, Fungicide, and
    Rodenticide Act (FIFRA; U.S.,
    1947, 1972), **2**:77, 101
Federal Mining Act, *see* Mining Law of
    1872
Federal Power Commission (FPC; U.S.),
    **2**:100
Federal Water Pollution Control Act, *see*
    Clean Water Act
feeding guilds, **2**:119*l*
feedlots, **1**:11; **3**:123
feminism, ecofeminism and, **2**:40
ferns, **2**:102-3
    coal formed from, **1**:105
ferrets, black-footed, **3**:103
ferrous metals, **4**:28
fertile organisms, **3**:68*l*
fertile soils, **6**:13, 14*l*
fertility rates, **2**:103
fertilization, of flowering plants,
    **2**:114-15
fertilizer minerals, **4**:28
fertilizers
    agricultural pollution resulting from,
      **1**:10-11
    global warming and, **3**:27
    Green Revolution and, **3**:39
    organic, **4**:101
    synthetic, **5**:111; **6**:14
fiber, **3**:57, 58*l*
fiber products, **2**:124
field studies, **2**:129*l*
    Fossey's, **2**:128-29
FIFRA, *see* Federal Insecticide,
    Fungicide, and Rodenticide Act
Finch, Robert, **2**:6-7
firebreaks, **5**:39-40
fire ecology, **2**:103-5
firelines, **2**:123*l*
fires
    forest fires, **2**:104-5, 122-23; **4**:56-57

    secondary succession following,
      **5**:129
    in Yellowstone National Park, **6**:74
    on prairies, **5**:35
    prescribed burns, **5**:38-40
    in savannas, **5**:90
fishes, **2**:105-7, 110*t*; **6**:30-31
    aquacultural growing of, **1**:34-35
    bacteria inside of, **1**:50
    Fish and Wildlife Service and, **2**:106-7
    gills of, **3**:20*l*
    salmon, **5**:84-86
    snail darter, **2**:62; **6**:6-7
    tuna, **6**:23
fish farming, **1**:35
fishing, **2**:110*t*
    commercial, **2**:107-9
      gill nets for, **3**:19-20
      legislation on, **5**:86
      sea turtles and, **5**:95, 96
    recreational, **2**:109-11
    for salmon, **5**:85
    territorial waters extended for, **5**:27
    for tuna, **6**:23
fish ladders, **2**:111-12; **4**:26; **5**:86
Fish and Wildlife Service (FWS; U.S.),
    **2**:106-7
    endangered species list maintained
      by, **2**:61, 62
    state hunting laws and, **6**:69
    wetlands under, **6**:56
    wildlife refuges of, **4**:36, 51-54; **5**:44
fixed nitrogen, **4**:67
flies, **5**:29
Flood Control Act (U.S., 1938), **1**:41
flood irrigation, **3**:98
floodplains, **2**:112-13
floods, **1**:41-42; **2**:112; **4**:56; **5**:73
floos, **6**:52
flora, *see* plants
Florida
    alligators in, **6**:67
    Everglades National Park in, **2**:35,
      86-88; **4**:44; **5**:88
    panthers in, **2**:58; **5**:38
    Pelican Island in, **4**:51
Florida panthers, **2**:58; **5**:38
flowering plants (angiosperms), **1**:109;
    **2**:113-15; **5**:21
    pollination of, **5**:28-29
    polypoidy among, **5**:122
flowers, **2**:114
fly ash, **2**:116; **5**:117
food chains, **2**:44, 116-17
    amphibians in, **1**:29
    aquatic, **5**:14
    autotrophs in, **1**:47

# G

methane released into atmosphere by decay of, **4**:21
waste management of, **6**:36
*see also* solid wastes
garbage dumps, **3**:105
garbologists, **3**:4*l*, 5
gasohol, **1**:47; **2**:134; **3**:7
gasoline
    gasohol and, **3**:7
    lead added to, **3**:116
    used by automobiles, **1**:46-47
        unleaded, **1**:90
Geiger, Robert, **2**:38
gene banks, **3**:10-11
gene pools, **3**:11-12
    hybridization and, **3**:68
generators, **4**:89*l*, 90
genes, **3**:7-10; **5**:119-21
    in bacteria, **1**:48, 50
    biodiversity and, **1**:58
    carcinogens and, **1**:80
    diversity within species of, **3**:12
    DNA in, **2**:29-32
    engineering of, **3**:12-15
    in evolution, **2**:90
    hybridization of, **3**:68-69
    in populations (gene pools), **3**:11-12
    resistance to pesticides in, **5**:2
    study of (genetics), **3**:15-16
genetic diseases, **3**:56
genetic diversity, **1**:58; **3**:9-10, 12
    Green Revolution and, **3**:39
    needed for natural selection, **4**:62
genetic drift, **5**:121
genetic engineering, **1**:69; **3**:12-15
genetics, **3**:15-16
    diseases linked to, **3**:56
    DNA in, **2**:29-32
    hybridization in, **3**:68-69
geology
    Darwin on, **2**:3
    of ice ages, **3**:76
    impact of glaciers on, **3**:23-24
    natural disasters
        earthquakes, **4**:54-55
        volcanic eruptions, **4**:55-56
    plate tectonics in, **5**:23-25
    of volcanism, **6**:34-35
geometric growth, **5**:34-35
    *see also* exponential growth
geothermal energy, **1**:26; **2**:52, 134; **3**:16-18
geysers, **3**:16, 17*l*
giant pandas, **3**:18-19
giant sequoia trees, **6**:75
gill nets, **3**:19-20
gills, **2**:106; **3**:20*l*

glaciation, **2**:79; **3**:20
glaciers, **2**:79; **3**:21-24; **5**:96
    effects of global warming on, **3**:27
    during ice ages, **3**:76
Global Environmental Monitoring System (GEMS), **6**:28
Global 2000 Report, **1**:130; **3**:25
global warming, **3**:25-28, 35-36
    carbon dioxide and, **1**:84-85
    deforestation and, **2**:15-16
    greenhouse effect and, **1**:85
    greenhouse gases and, **3**:27*c*, 37
    impact on tundras of, **6**:25
Global Warming Convention (1992), **6**:28
Goodall, Jane, **2**:128
gorillas, **3**:28-29
    mountain gorillas, **2**:128-29
        poaching of, **5**:26
    parasites of, **4**:116
Grant, Ulysses S., **2**:101; **3**:112; **4**:30, 43, 46; **5**:42
grasses
    on prairies, **5**:35, 36
    prescribed burns of, **5**:39
    in savannas, **5**:89-91
grasslands, **1**:67; **3**:29-32
    fire dependency of, **2**:104
    habitats in, **3**:46-47
    legislation on grazing on, **6**:5
    national grasslands, **2**:127-28; **4**:40-41
    overgrazing of, **4**:103
    prairies, **5**:35-37
    prescribed burns in, **5**:38-39
    rangelands and, **5**:57
    savannas, **5**:89-91
gravity, **2**:81*l*
    erosion caused by, **2**:79
grazing, **3**:32-33, 122-23
    legislation on, **6**:5
    in national grasslands, **4**:40-41
    overgrazing and, **4**:103-4
    on prairies, **5**:36
    on public lands, **2**:128
    on rangelands, **5**:57-58
    on savannas, **5**:90
Great Ice Age, **3**:76
Great Lakes (U.S.-Canada), **3**:33-34, 76; **6**:47
green algae, **1**:23; **5**:17
greenhouse effect, **1**:85, 102; **3**:34-36
    agroforestry to prevent, **1**:16
    deforestation and, **2**:15-16
    global warming and, **3**:25
    greenhouse gases in, **3**:36-37, 36*t*
greenhouse gases, **1**:101-2; **2**:122; **3**:26-27, 27*c*, 35-37, 36*t*

carbon dioxide, **1**:82-85
methane, **4**:21-22
natural gas, **4**:58
nitrogen monoxide, **1**:17
solar heating and, **5**:113
green-manure plants, **5**:111
Green Mountain National Forest (Vermont), **4**:40
green politics, **3**:37
Green Revolution, **3**:38-39
grizzly bears, **3**:39-41; **4**:98; **5**:38, **6**:74
ground moraines, **3**:23-24
groundwater, **1**:35; **3**:97-98; **5**:82, 83*l*, 134; **6**:49
    pollution of, **1**:37
        leachates in, **1**:56; **3**:106
        from mining, **4**:30
        from pesticides, **5**:1
        restoration of, **1**:38
        from toxic wastes, **6**:15-17
    saltwater intrusion into, **5**:88-89
    used in geothermal energy, **3**:16-18
    in zone of saturation, **6**:76
Gulf Stream, **4**:86
gullying, **3**:41-42; **5**:108
gymnosperms (plants), **5**:21
gypsy moths, **3**:42-43

# ▶ H

habitats, **2**:42; **3**:44-47
    adaptive radiation in, **1**:8
    impact of dams on, **2**:2-3
    loss of, **3**:47-49, 47*g*
        for California condors, **1**:79
        for endangered species, **2**:58
        extinction caused by, **1**:74; **2**:94
        mining and, **4**:29
        for mountain gorillas, **2**:128
        for northern spotted owls, **4**:73
        for passenger pigeons, **4**:118
        population growth and, **5**:34
        for predators, **5**:38
        for sea turtles, **5**:95
        for tigers, **6**:12
    management of, **6**:69
    migrations and, **4**:25, 26
    old-growth forests as, **4**:97
    in rain forests, **5**:54-55
    wetlands, **6**:54-56
    wilderness, **6**:61
    wildlife in, **6**:63-65
        conservation of, **6**:65-68
        refuges for preservation of, **4**:53; **5**:44

Haeckel, Ernst, **2**:41
Hahn, Otto, **4**:75
hail, **6**:54
Hanford (Washington), **3**:49
haploid plants, **5**:19
Hardin, Garrett, **3**:49-50; **6**:18
hard water, **2**:27
hardwood, **2**:120*l*
harvesting, **2**:108*l*
  of fish, **2**:109
hatcheries, **5**:84*l*, 86
Hawaii
  birds of, **2**:92
    Hawaiian geese (nene), **6**:78
   mongooses introduced in, **5**:2
Hawaiian geese (nene), **6**:78
Hazardous Materials Control Act (U.S.,
    1970), **3**:50
Hazardous Materials Transportation Act
    (HMTA; U.S., 1975) **3**:50-51
Hazardous Materials Transportation
    Uniform Safety Act (U.S., 1990),
    **3**:50
hazardous substances, **1**:113; **3**:50-52
Hazardous Substances Act (U.S., 1960),
    **3**:51-52
hazardous wastes, **3**:52-53; **6**:15-17
  disposal of, **3**:52*c*, 80-82; **6**:37-38
  legislation on, **5**:69-70, 115, 129-32
  in Love Canal, **3**:123-24
  management of, **6**:36
  radioactive, **5**:48-50
  storage and transportation of,
    **3**:53-55
  in water supplies, **5**:82
  *see also* toxic wastes
hazards, **3**:51*l*
  of dumping, **3**:105
  to health, **3**:56-57
  of landfills, **3**:106
health, **3**:55-57
  diseases and, **3**:55-57
  environmental law on, **3**:113
  insects and, **3**:87
  malnutrition and, **4**:2-4
  medical wastes and, **4**:17-18
  noise pollution and, **4**:71
  nutrition and, **3**:57-60
  problems linked to mining, **4**:29
  radiation exposure and, **5**:46-47, 47*t*
  sick building syndrome and, **5**:102-3
  ultraviolet light and, **6**:26-27
  *see also* diseases
heat
  as marine pollution, **4**:11
  radiation of, **5**:46
  in thermal pollution, **5**:31, **6**:8

heating
  fuel wood for, **2**:134
  solar, **5**:113-14; **6**:5-6
heavy metals
  lead, **3**:115-17
  mercury, **4**:19
  poisoning from, **3**:61-62, 61*t*
    of fish and shellfish, **4**:9
    Minamata disease (mercury), **4**:27
  in tailings, **6**:4
hemoglobin, **3**:16*l*, 75
herbaceous plants, **1**:67, 67*l*
herbicides, **3**:62-64, 62*t*
  defoliants, **2**:12-13
  dioxins in, **2**:28
  legislation on, **2**:101
herbivores, **1**:29, 118, 120; **3**:64; **6**:20
  in food chains, **2**:116-17
  insects as, **3**:84
  grazing by, **3**:32-33
herb layer (in forests), **2**:120
Hershey, Alfred D., **2**:30
Hess, Harry, **5**:24
Hetch-Hetchy Valley (California), **5**:16
heterotrophs, **1**:118
  *see also* consumers
hibernation, **6**:63, 64*l*
high-grade ores, **6**:3
high-level radioactive wastes, **5**:49-50,
    91
Hippocrates, **1**:80
histograms, **1**:9*g*, 10
Historic Sites Act (U.S., 1935), **4**:44
homeostasis, **3**:2
home ranges, **3**:28*l*
*Homo sapiens* (humans), **5**:119
  *see also* humans
Hooker Chemical Company, **3**:52
Hoover Dam, **1**:76-77
hormones, **1**:86*l*
horned lizards, **5**:67
hosts, **1**:62*l*; **4**:115*l*; **5**:138
houses, *see* buildings
Howard, Sir Albert, **4**:99
humans
  animals and
    dolphins and porpoises, **2**:34-35
    insects, **3**:86-87
    migrations of, **4**:25-26
    pets, **5**:6-8
    seals and sea lions, **5**:93-95
    sea turtles, **5**:95
    tiger attacks on, **6**:13
    whales, **6**:58-59
  bacteria and, **1**:50; **5**:138
  in biophilia hypothesis, **1**:68
  carbon cycle and, **1**:83-84

as consumers, **1**:119
coral reefs and, **1**:128
drinking water for, **6**:42-43
ecosystems restored by, **5**:71-72
ecotourism by, **2**:48
exotic species introduced by, **2**:91
extinctions caused by, **1**:59; **2**:95-96;
    **4**:15
family planning for, **2**:98-99
  contraception use and, **2**:98*g*
forest fires caused by, **2**:122-23
in frontier ethic, **2**:131-32
habitat loss caused by, **3**:47-48
health of
  causes of death of, **3**:55*c*
  diseases and, **3**:55-57
  effects of pesticides on, **4**:122
  genetic diseases of, **3**:14
  infant mortality and, **3**:82-83
  lead poisoning of, **2**:75
  malnutrition and, **4**:2-4
  nutrition and, **3**:57-60
  pathogens and, **4**:118
  radiation exposure and, **5**:47, 47*t*
  sick building syndrome and, **5**:102-3
  *see also* diseases
history of
  agricultural revolution in, **1**:13-14
  evolutionary origins of, **1**:69; **5**:36
  Industrial Revolution in, **3**:78-80
  of petroleum use, **5**:4
  population growth in, **5**:34
  sewage treatment in, **5**:99
  smelting in, **5**:105
  subsistence agriculture in, **5**:125
impacts of
  on biosphere, **1**:72
  on climate, **1**:101-2
  on nitrogen cycle, **4**:68
  on species diversity, **5**:123
  on wildlife, **6**:64
importance of forests to, **2**:122
importance of grasslands to, **3**:30-32
land use by, **3**:110-12
methane released into atmosphere
    by, **4**:21
oceans and, **4**:84-85
as omnivores, **4**:98
ozone and, **4**:113
populations of, **1**:xi
  demography of, **2**:16-17
  Global 2000 Report on, **3**:25
  limiting factors on, **3**:121
  overpopulation and, **4**:104
primary pollution by, **5**:40
rain forests and, **5**:55-56
societies of

 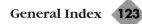

of oil shale, **4**:94
open-pit, **4**:98-99
radioactive wastes from, **5**:48
solid wastes from, **5**:115
  management of, **6**:36
  source reduction of, **5**:117-18
spoil piles from, **3**:115
strip mining, **5**:124-25
surface mining, **5**:132-33
tailings from, **6**:3-5
Mining Law of 1872 (Federal Mining Act;
    U.S.), **2**:101-2; **4**:30-31
Mining and Mineral Policy Act (U.S.,
    1970), **4**:28
Minnesota Mining and Manufacturing
    Company, **6**:39
minorities, environmental justice and,
    **2**:75
miracle rice, **3**:38
Mississippi River, watershed of, **6**:50
mistletoe, **5**:138
monerans, **5**:119
mongooses, **2**:92; **5**:2
monocotyledons (monocots; plants),
    **2**:113
monoculture farming, **1**:97; **4**:31-32, 121;
    **5**:2
  tree farming as, **6**:19
Mono Lake Committee, **5**:45
monomers, **5**:21
montane forests, **1**:115-16
Montreal Protocol (1987), on CFCs, **1**:19;
    **4**:32, 111
moose, **6**:2-3
moraines, **3**:20, 23-24
Morgan, Ann Haven, **4**:33
motor vehicles, *see* automobiles
mountain gorillas, **2**:128-29; **5**:26
mountains, plate tectonics and formation
    of, **5**:25
Mount Pinatubo (Philippines), **6**:35
Mount Saint Helens (Washington), **6**:35
movements, *see* social and political
    movements
muck, **5**:99*l*
Muir, John, **4**:33-34; **5**:16; **6**:75
mulch, **2**:81
mules, **3**:68
Muller, Paul, **2**:5
multiple chemical sensitivity (MCS),
    **5**:102
multiple use of public lands, **2**:126;
    **4**:34-36
  debate over, **6**:63
  of forests, **4**:35*c*, 39
  of national resource lands, **5**:44
  of wildlife refuges, **4**:53-54

Multiple Use-Sustained Yield Act (U.S.,
    1960), **4**:34-36
mummification, **3**:4*l*, 5
municipal solid wastes, **5**:117-18
muscular systems, **6**:30
museums, natural history, **4**:58-59
mushrooms, **2**:138
mutants, **1**:48, 49*l*
mutations, **3**:9, 12, 15
  caused by carcinogens, **1**:80
  errors in DNA copying leading to,
    **2**:31
  evolution and, **2**:90
mutualism, **1**:63; **4**:36-37; **5**:136
mycologists, **2**:136

# ▶ N

Naess, Arne, **2**:11, 12
nanometers, **5**:46*l*
National Ambient Air Quality Standard
    (NAAQS), **1**:53, **2**:77
National Conservation Commission
    (U.S.), **5**:80
National Consortium for Environmental
    Education and Training (NCEET),
    **2**:70
National Environmental Policy Act
    (NEPA, U.S.), **1**:129; **4**:38
  environmental impact statements
    under, **2**:71-74
  Environmental Protection Agency
    enforcement of, **2**:77
national forests, **2**:126; **4**:38-40; **5**:44-45
  Forest Service and, **2**:127-28
  multiple use of, **4**:34-36
  tree farming on, **6**:18-19
national grasslands, **2**:127-28; **4**:40-41;
    **5**:37; **6**:5
National Marine Fisheries Service (U.S.),
    **2**:62
national monuments, **5**:43
National Museum of Natural History
    (Washington, D.C.), **4**:59
National Oceanic and Atmospheric
    Administration (NOAA; U.S.),
    **4**:41-42
  National Ocean Service of, **4**:12
  National Weather Service of, **4**:48-51;
    **6**:54
National Ocean Service (U.S.), **4**:12
national parks, **4**:42-46; **5**:43; **6**:66
  Everglades National Park, **2**:86-88
  mining in, **4**:30
  National Park Service and, **4**:46-47

Yellowstone National Park, **4**:34; **6**:74
  Yosemite National Park, **6**:75
National Park Service (U.S.), **2**:73; **4**:36,
    43-47; **5**:43; **6**:66
  national seashores under, **4**:48
  policy on fires of, **2**:104-5
National Park Service Act (U.S., 1916),
    **4**:43, 46
National Park System (U.S.), **4**:43, 44,
    46-47
National Pollutant Discharge Elimination
    System (NPDES), **4**:47-48; **6**:47
national resource lands, **5**:44
national seashores, **4**:48
National Weather Service (U.S.), **4**:42,
    48-51; **6**:54
National Wilderness Preservation System
    (NWPS; U.S.), **5**:42-43; **6**:61
national wildlife refuges (U.S.), **4**:51-54;
    **5**:44, **6**:65, 69
  Arctic National Wildlife Refuge, **1**:21,
    40-41
  Yellowstone National Park as, **6**:66
National Wildlife Refuge System (U.S.),
    **4**:51-54; **6**:69
Native Americans (American Indians),
    **3**:78
  crude oil used by, **5**:4
  Interior Department and, **2**:19, 20
  *see also* indigenous peoples
native species, **2**:91; **4**:54
  protecting, **6**:68-69
  threatened by exotic species, **2**:92
natural disasters, **4**:54-57
  famines and, **2**:99
  volcanoes, **6**:35
natural gas, **2**:131, 132; **4**:57-58; **5**:4
  Federal Energy Regulatory
    Commission and, **2**:100-101
  methane in, **1**:19; **4**:21
  reserves of, **4**:57*c*
  reservoirs of, **4**:92
  synthetic (SNG), **5**:139
natural history, **3**:28*l*, 29
natural history museums, **4**:58-59
natural pollutants, **6**:46-47
natural resources, **4**:59-61
  conservation of, **1**:116-18
    history of, **1**:xi
    legislation on, **5**:69-70
  ecological economics of, **2**:40
  frontier ethic on, **2**:131
  Interior Department responsibilities
    for, **2**:19
  International Union for the
    Conservation of Nature and
    Natural Resources, **3**:91-92

land use and, **3**:110-12
laws protecting, **3**:112-13
minerals, **4**:28, 28*c*
nonrenewable, **4**:72
preserves for, **3**:108*l*
public versus private ownership of,
     **6**:18
in rain forests, **5**:55
recycling, reducing, and reusing,
     **5**:59-62
renewable, **5**:62-65
sustainable development of, **5**:135-36
natural selection, **1**:8; **2**:90-91; **3**:12, 16;
     **4**:61-62; **5**:120, 122
     Darwin on, **2**:3, 4
     within gene pools, **3**:11
     in resistance to pesticides, **4**:123
     resistance to pesticides by, **5**:2
Nature Conservancy (TNC), **3**:108-9;
     **5**:37, 39
Nelson, Gaylord, **2**:68
nene (Hawaiian geese), **6**:78
neritic zone (of oceans), **4**:84
Nevada, Yucca Mountain in, **5**:50
New Source Performance Standards
     (NSPS), **1**:53
newts, **1**:28
New York (state), Love Canal in, **1**:93;
     **3**:52-53, 80, 123-24; **5**:131; **6**:17
New Zealand, **2**:59-60
niches, **2**:42; **4**:62-65; **6**:20
NIMBY syndrome (not in my backyard),
     **3**:108; **4**:65-67
nitrogen
     in atmosphere, **1**:43
     in legumes, **3**:117
nitrogen cycle, **1**:47; **4**:67-68
nitrogen dioxide, **1**:17-18, 107; **4**:68-69
nitrogen fixing, **1**:49; **4**:67-69, 108
nitrogen monoxide, **1**:17
nitrogen oxides, **1**:17-18, 90; **3**:27;
     **4**:69-70; **5**:41
     as greenhouse gases, **3**:35, 36
     nitrogen dioxide, **4**:68-69
     nitrogen monoxide, **1**:17
     in smog, **5**:106
Nixon, Richard M., **2**:71, 76
Noise Control Act (U.S., 1972), **4**:71
noise pollution, **4**:70-71, 70*c*; **5**:30-31
nonferrous metals, **4**:28
nonmetallic minerals, **4**:28
nonnative species, **1**:58*l*
     *see also* exotic species
nonpoint sources (of pollution), **4**:71-72;
     **5**:27
     of water pollution, **6**:45, 47
     in oceans, **4**:85

of surface water, **5**:134
of wastewater, **6**:40
nonrenewable resources, **1**:116; **4**:61, 72;
     **5**:59, 63
     coal as, **1**:106
     conservation of, **1**:117-18
     fossil fuels as, **2**:129
     ocean thermal energy as, **4**:90
nonvascular plants, **5**:20
North America
     mammals in, **4**:6
     prairies of, **5**:35-37
North American Waterfowl Management
     Plan (NAWMP), **6**:56
North American Wetlands Conservation
     Act (U.S., 1990), **6**:56
North Carolina
     Cape Hatteras National Seashore in,
          **4**:48
     Pigeon River in, **5**:27-28
northern spotted owls, **1**:98; **3**:49;
     **4**:72-74, 97; **6**:19, 65, 67
no-till agriculture, **1**:12; **4**:74-75; **5**:110
nuclear fission, **4**:75-76
     breeder reactors and, **1**:75
     for nuclear weapons, **4**:80
     uranium used in, **6**:29
nuclear fusion, **4**:76-77, 80
nuclear power, **2**:50, 133; **4**:78-79
     Atomic Energy Commission and, **1**:45
     breeder reactors for, **1**:75
     Chernobyl accident and, **1**:93-94
     environmental problems associated
          with, **2**:55
     International Atomic Energy Agency
          and, **3**:89-90
     nuclear fission and, **4**:75-76
     nuclear fusion and, **4**:76-77
     plutonium and, **5**:25-26
     radioactive wastes from, **5**:33
     wastes from
          at Hanford, **3**:49
          radioactive, **5**:48-50
nuclear power plants, **4**:78-79
     core meltdowns in, **6**:11*l*
     radioactive fallout from accidents at,
          **5**:48
     radioactive wastes from, **5**:48-50
     Three Mile Island, **6**:10-11
nuclear reactors, *see* nuclear power
     plants
Nuclear Regulatory Commission (NRC;
     U.S.), **1**:45
nuclear wastes
     at Hanford, **3**:49
     seabed disposal of, **5**:91
nuclear weapons, **4**:80-81

Hanford facility for, **3**:49
nuclear fission and, **4**:75-76
nuclear fusion and, **4**:77
nuclear winter as result of, **4**:81
plutonium used in, **5**:26
radioactive fallout from, **5**:47-48
United Nations Treaty on the Non-
     Proliferation of Nuclear Weapons,
     **3**:90
nuclear winter, **4**:81
nucleus (of cell), **1**:48, 49*l*
nutrients, **3**:46*l*, 57-59; **4**:2
     bacterial action on, **1**:49
     cycling within ecosystems of, **2**:44-45
     needed by phytoplankton, **5**:14-15
     in oceans, **3**:44; **4**:86
     in soils, **4**:31-32
     in topsoil, **6**:14
nutrition, **3**:57-60
     malnutrition and, **4**:2-4
     vitamin and mineral deficiency
          diseases, **4**:3*t*

# ▶ O

Occupational Safety and Health
     Administration (OSHA; U.S.), **4**:71
ocean currents, **1**:100; **4**:85-87
     El Niño, **2**:57
     estuaries formed by, **2**:81
     plankton carried in, **5**:17
ocean dumping, **3**:105; **4**:87-89
     legislation on, **4**:11-12
     whales harmed by, **6**:59
Ocean Dumping Ban Act (U.S., 1988),
     **4**:85, 89
oceanic zone, **4**:84
oceanographers, **4**:82
oceanography, **4**:82
ocean pollution, **1**:130
oceans, **3**:74; **4**:82-85
     anadromous species in, **5**:84
     coral reefs in, **1**:126-28
     currents in, **1**:100; **4**:85-87
     desalinization of water from, **2**:20-21
     dredging of, **2**:35-36
     effects of global warming on, **3**:27
     intertidal zones and, **3**:92-95
     Law of the Sea Convention on,
          **3**:114-15
     legislation protecting, **4**:11-12
     as marine habitats, **3**:44
     National Oceanic and Atmospheric
          Administration and, **4**:41-42
     petroleum under, **4**:84*c*

phytoplankton in, **5**:14
plankton in, **5**:17-18
poaching in, **5**:27
pollution in, **4**:9-11; **6**:45
  oil, **4**:93-96; **5**:6
salinity of waters in, **5**:83
seabed disposal in, **5**:91-92
sea-floor spreading under, **5**:24
sediment in, **5**:96
tidal energy from, **6**:11-12
wastes dumped into, **4**:87-89
in water cycle, **6**:43
zooplankton in, **6**:78
ocean thermal energy, **1**:26-27; **4**:89-90
ocean thermal energy conversion
  (OTEC), **4**:89-90
ocean trenches, **4**:83
ocean waves, **2**:79-80
Odum, Eugene Pleasants, **4**:90-91
Odum, Howard T., **4**:90, 91
Oersted, Hans Christian, **2**:53
Office of..., *see under main part of name*
offspring, **3**:68*l*
oil, *see* petroleum
oil bilge washings, **4**:95*l*, 96
oil drilling, **4**:91-93
oil pollution, **4**:93
oil shale, **4**:94
oil spills, **4**:94-96; **5**:6
  in Alaska (1989), **1**:21; **2**:96
  in Antarctica, **1**:34
  marine pollution from, **4**:11
  microbes use to clean up, **1**:69-70
  on permafrost, **1**:41
  during Persian Gulf war, **1**:92
  whales harmed by, **6**:59
old-growth forests, **4**:39, 97
  northern spotted owls in, **4**:72-74
omnivores, **1**:29, 119; **4**:98
open-pit mining, **4**:29, 98-99; **5**:132
Ordovician mass extinction, **4**:13
Oregon, air quality legislation in, **1**:20
ores, **5**:105; **6**:4*l*
organic compounds, **5**:108
organic farming, **4**:99-101
  subsistence agriculture and, **5**:126
  sustainable agriculture and, **5**:135
organic fertilizers, **4**:101
Organic Foods Production Act (U.S.,
  1990), **4**:100
organic pollutants, **6**:45-46
organic wastes
  decomposition of, **2**:10, 11
  detritus, **2**:28
  septic tanks for, **5**:97-98
  in sewage, **5**:99-100
  treatment of, **6**:41

organisms, **1**:58*l*
  adaptation by, **1**:6-8
  aerobic, **1**:8
  bacteria inside of, **1**:50
  biological communities of, **1**:61-63
  in biosphere, **1**:71
  in buildings, **5**:102
  coevolution of, **1**:108-9
  consumers, **1**:118-20
  in deciduous forests, **2**:9-10
  decomposers, **2**:10-11
  on deserts, **2**:23-24
  diversity of, **2**:41*l*
  in energy pyramids, **2**:65-66
  evolution of, **2**:89-91
  Gaia hypothesis on, **3**:2
  habitats of, **3**:44-47
    loss of, **3**:47-49
  interactions among, **2**:42-44
  in intertidal zones, **3**:94-95
  microscopic, **5**:13, 14*l*
  native species, **4**:54
  parasitic, **4**:122*l*
  parasitism among, **4**:115-16
  predators, **5**:37-38
  as producers, **5**:41
  respiration of, **5**:70-71
  in soil, **5**:108
  species of, **5**:118-22
  symbiosis among, **5**:136-38
  wildlife, **6**:63-65
  *see also* species
Organization of Petroleum Exporting
  Countries (OPEC), **4**:101-2
organizations
  environmental, **6**:84-94
  governmental (U.S.), **6**:95-96
  *see also* environmental organizations
orphaned wildlife, **6**:69-70
Outer Banks (North Carolina), **4**:48
overburden, **5**:124, 132
overgrazing, **3**:32-33, 122-23; **4**:103-4
  legislation on, **6**:5
  of rangelands, **5**:58
  of savannas, **5**:90-91
overharvesting, **3**:34*l*
overpopulation, **4**:104
overproduction, **2**:89
owls, northern spotted, **1**:98; **3**:49;
  **4**:72-74, 97; **6**:19, 65, 67
oxisols, **5**:109
oxpecker birds, **5**:136
oxygen, **4**:105-6
  aerobic organisms needing, **1**:8
  anaerobic organisms not needing,
    **1**:30
  in atmosphere, **1**:44

biochemical oxygen demand, **1**:55-56
dissolved, **2**:29
  in aquatic ecosystems, **1**:11
  eutrophication and, **2**:85
  hypoxia (deficiency of) and, **3**:75
  organic pollutants and, **6**:46
  in oxygen cycle, **4**:107-8
  ozone form of, **4**:108-9
    ozone hole, **4**:109-11
    ozone layer, **4**:111-12
    as pollutant, **4**:112-13
    in stratosphere, **5**:124
  photosynthesis and, **5**:12
oxygen cycle, **4**:107-8
oxygen-demanding wastes, **6**:46
ozone, **1**:91; **4**:106, 108-9
  ground-level, **1**:18
  in ozone layer, **4**:111-12
  as pollutant, **4**:112-13
  in stratosphere, **1**:44-45; **5**:124
ozone hole, **1**:91; **4**:109-11
ozone layer, **1**:44-45; **4**:106, 108, 111-12
  chlorofluorocarbons in, **1**:9, 90-91
  Freons in, **2**:131
  hole in, **4**:20, 109-11
  Montreal Protocol on, **4**:32
  ultraviolet radiation and, **2**:55; **6**:27

# ▶ P

packaging materials, **6**:39
pack ice, **1**:38
paleontology, **4**:58, 59*l*
pampas, **4**:114-15
pampas grass, **4**:114-15
pandas, **3**:18-19
Pangaea, **1**:121; **5**:23
panthers, **2**:58; **5**:38
parasites, **1**:87; **4**:115-16, 115*l*, 122*l*;
  **5**:138
  in biological control systems, **1**:64
  in biological pesticides, **4**:124
  fungi as, **2**:138
  insects as, **3**:85-86
parasitism, **1**:63; **2**:43-44; **4**:115-16, **5**:138
Park, Thomas, **1**:111
particulates, **1**:18; **4**:116-17; **5**:30
  scrubbers to reduce, **5**:91
  volcanic ash, **5**:40-41
passenger pigeons, **4**:25, 117-18; **6**:65
passive solar heating, **5**:112, 114
pathogens, **3**:56; **4**:118
  in biological control systems, **1**:64-65
  insects as vectors of, **3**:87
  in wastewater, **6**:39

Patrick, Ruth, **4**:118-19
PCBs (polychloronated biphenyls), **4**:119-20; **6**:15
peat, **1**:105-6; **2**:130; **4**:120-21
peat bogs, **4**:121
Pelican Island (Florida), **4**:51
Pennsylvania, Three Mile Island Nuclear Power plant in, **4**:65-66, 79; **6**:10-11
permafrost, **1**:38; **3**:47, 75; **4**:21*l*; **6**:24
oil spills on, **1**:41
permeable soils, **5**:76*l*, 77
Permian mass extinction, **4**:14
Persian Gulf war, **1**:92; **4**:94
Peru, pampas of, **4**:114-15
pest control, **4**:121-24
integrated pest management for, **3**:89; **4**:99, 101
pesticides, **4**:121-24; **5**:1*t*, 1-2
as agricultural pollution, **1**:11-12
agroecology and, **1**:14
Carson's warnings on, **1**:88-90
DDT, **2**:5-7
fungicides, **2**:138-39
Green Revolution and, **3**:39
in groundwater, **1**:37
herbicides, **3**:62-64, 62*t*
impact of food webs of, **2**:119
insecticides, **3**:88, 88*t*
integrated pest management alternative to, **3**:89
legislation on, **2**:77, 101
rodenticides, **5**:78-79
synthetic, **1**:63
used on gypsy moths, **3**:43
*see also* DDT
petrochemicals, **5**:3
plastics and, **5**:22
petroleum, **2**:130-32; **5**:3-6
Alaska Pipeline for, **1**:20-21
under Arctic National Wildlife Refuge, **1**:40-41; **4**:54
drilling for, **4**:91-93
under oceans, **4**:84*c*
oil shale and, **4**:94
Organization of Petroleum Exporting Countries and, **4**:101-2
petrochemicals and, **5**:3
pollution from, **4**:93
used in making plastics, **5**:22
*see also* gasoline; oil spills
petroleum industry, **5**:5-6
deep-well injection used by, **6**:38
pet trade, **5**:6-8
Philippines, **6**:35
phloem (in plants), **5**:21
phosphates, **2**:27; **5**:9-10

algal blooms and, **2**:85
in sewage, **5**:100
strip mining of, **5**:124
phosphorus, **3**:34*l*; **5**:9-10
in algal blooms, **1**:24
photochemical oxidants, **1**:95*l*
photochemical smog, **1**:107; **5**:106
photodegradable plastics, **5**:10-11, 23
photosynthesis, **1**:2, 7, 60; **5**:11-12
in algae, **1**:22
in carbon cycle, **1**:83
by cyanobacteria, **1**:48
deforestation and, **2**:16
greenhouse effect and, **3**:36
in oxygen cycle, **4**:105, 108
by phytoplankton, **5**:14
by plankton, **5**:17
by plants, **5**:18-19
by producers, **1**:47; **5**:41
in rain forests, **5**:56
respiration as opposite of, **5**:71
solar radiation and, **5**:46
photovoltaic (PV) cells, **1**:25-26; **2**:50; **5**:13, 112; **6**:6
photovoltaic systems (PV), **1**:25
pH values, **1**:4*c*; **5**:9
phyla, **3**:96*l*
of invertebrates, **3**:95-96
physical control of pests, **4**:101, 124
physical hazards, **3**:57
physiological adaptations, **1**:6-7
phytoplankton, **1**:22; **4**:86; **5**:13-15, 17
killed by ultraviolet radiation, **6**:27
zooplankton distinguished from, **6**:78
Pigeon River (North Carolina), **5**:27-28
pigments, **1**:21*l*
Pinchot, Gifford, **2**:127; **4**:34, 40; **5**:15-16
pine trees, **2**:104
pioneer species, **2**:45; **5**:16-17, 128
Pisaster (starfish), **3**:102, 103
placental mammals, **1**:125; **4**:6
plankton, **4**:108; **5**:17-18
petroleum formed from, **5**:3
zooplankton, **6**:78
*see also* phytoplankton
plantations, **2**:13, 14*l*
plant pathology, **5**:18
plants, **5**:18-21, 119
acid rain's effects on, **1**:5
agroforestry and, **1**:15-16
annuals, **2**:22*l*, 23
in Arctic, **1**:39
as autotrophs, **1**:47
biomass of, **1**:65
birds' help to, **1**:74
botanists and, **4**:22*l*
bromeliads, **1**:2*l*

captive propagation of, **1**:81-82
carnivorous, **1**:87
coal formed from, **1**:105
coevolution with other organisms and, **1**:109
commensalism among, **1**:110
composting of, **1**:112
crop rotation of, **1**:131
diseases of
caused by fungi, **2**:138-39
pathology of, **5**:18
viruses, **6**:34
eaten by herbivores, **3**:64
eaten by insects, **3**:84
endangered and threatened, **2**:58-61; **6**:67
in estuaries, **2**:83
euphorbia, **1**:67*l*
in Everglades, **2**:88
evolution of, **2**:89-91
ferns, **2**:102-3
fire ecology and, **2**:103-4
flowering, **2**:113-15
gene banks for, **3**:10-11
of grasslands, **3**:29-30
green-manure, **5**:111
herbaceous, **1**:67*l*
herbicides used against, **3**:62*t*, 62-64
irrigation of, **3**:97-100
legumes, **3**:117
medicines from, **1**:59*t*, 60
monoculture farming of, **4**:31-32
in nitrogen cycle, **4**:68
nitrogen-fixing bacteria inside of, **1**:50
in oxygen cycle, **4**:108
pampas grass, **4**:114-15
peat formed from, **4**:120
pest control for, **4**:121-22
pesticides produced by, **1**:65
petroleum formed from, **5**:3
phosphorus in, **5**:9
photosynthesis by, **5**:11-12
pollination of, **5**:28-29
polypoidy among, **5**:122
as predators, **5**:37
in rain forests, **5**:54; **6**:22
respiration by, **5**:71
in riparian lands, **5**:73
in salt marshes, **5**:86
salt tolerance of, **5**:84*t*
in savannas, **5**:89-91
sea oats, **2**:81*l*
succulents, **2**:22*l*, 23
symbiosis among fungi and, **5**:138
in taigas, **6**:2-3
trees

in water cycle, **6**:43
predation, **1**:62-63; **2**:43
predators, **1**:119; **2**:117; **5**:37-38; **6**:21
   in biological control systems, **1**:64
   exotic species as, **5**:2
   insects as, **3**:85
   for tigers, **6**:12-13
   whales as, **6**:59
prescribed burns, **2**:103, 123; **5**:38-40
preserves
   for natural resources, **3**:108*l*
   for wild animals, **1**:32*l*
      Antarctica as, **1**:33
prey animals, **3**:85; **5**:37
Priestley, Joseph, **5**:11, 12
primary pollution, **5**:40-41, 41*t*
primary succession, **5**:127-28
primary treatment
   of sewage, **5**:99, 101
   of wastewater, **6**:40-41
   of water, **6**:51
primates, **2**:34, 35*l*, 129*l*
   gorillas, **2**:128-29; **3**:28-29
private wells, **5**:83
producers (organisms), **2**:42, 116-17;
      **5**:41, 63; **6**:20
   algae as, **1**:22
   consumers distinguished from, **1**:118
   in energy pyramids, **2**:65-66
   phytoplankton as, **5**:14
   *see also* autotrophs
products
   agricultural, **3**:34*l*
   from algae, **1**:23
   biodegradable, **1**:56-57
   bottled water, **5**:82
   energy efficiency of, **2**:63-64
   of forests, **2**:123-25
   labeling of
     environmental, **3**:104
     of foods, **2**:18; **3**:60
   life-cycle assessments of, **3**:119-20
   packaging materials for, **6**:39
   petrochemical, **5**:3
   plastics in, **5**:22
   reusing, **5**:61-62
   source reduction of, **5**:118
   *see also* foods
Progressive Animal Welfare Society
     (PAWS), **1**:31
propellants, **4**:111, 112*l*
proteins, **3**:57; **6**:34*l*
protists, **1**:21*l*; **5**:119
protocooperative relationships, **5**:66
protozoans, **3**:96*l*; **5**:17
public lands, **5**:42-45
   Bureau of Land Management for, **1**:76

debate over use of, **6**:63
   grazing on, **6**:5
   mining on
     legislation on, **2**:101-2; **4**:28, 30-31
     reclamation of, **4**:91
   multiple use of, **4**:34-36
     of forests, **4**:35*c*
   national forests, **2**:126-28; **4**:38-40
   national grasslands, **4**:40-41; **5**:37
   national parks, **4**:42-46; **6**:66
   national seashores, **4**:48
   national wildlife refuges, **4**:51-54;
     **6**:65, 69
   overuse of, **6**:18
   Reclamation Act and, **5**:59, 80
   wilderness, **6**:60-62
Public Rangelands Improvement Act
     (U.S., 1978), **3**:33; **5**:58
public trusts, **5**:45
pulp, **2**:125, 126*l*
purse seining, **6**:23

# R

rabies, **5**:78*l*
radiation, **5**:32-33, 46
   as carcinogen, **1**:86-87
   electromagnetic spectrum of, **2**:54-55,
     54*c*
   exposure to, **5**:46-47, 47*t*
   from nuclear energy, **2**:133
   radioactivity and, **5**:51-52
   solar, **5**:113
   at Three Mile Island, **6**:11
   as treatment for cancer, **1**:81
   ultraviolet, **6**:26-27
   x rays as, **6**:73
   *see also* ultraviolet radiation
radioactive decay, **5**:51
radioactive fallout, **5**:47-48
radioactive materials, **6**:10, 11*l*
radioactive wastes, **5**:33, 48-50
   at Chernobyl accident, **1**:93-94
   marine pollution from, **4**:11
   from nuclear power plants, **4**:78
   seabed disposal of, **5**:91-92
   storage of, **3**:53-55
radioactivity, **4**:76; **5**:32-33, 51-52
   environmental problems associated
     with, **2**:55
   of plutonium, **5**:25
   as radiation, **5**:46
   of uranium, **6**:29
radio waves, **2**:54
radon, **5**:52-53, 103

*Rafflesia* (plant), **1**:109
rain, **5**:37; **6**:53-54
   acid rain, **1**:4-6
   deserts and, **2**:22
   on prairies, **5**:35, 36
   in river basins, **5**:77
   in savannas, **5**:89, 90
rain forests, **1**:68; **2**:121; **5**:53-56; **6**:21
   biodiversity in, **1**:58; **6**:22
   clear-cutting of, **1**:98
   conservation of, **1**:116-17
   deforestation in, **2**:13
   species in, **3**:48
rain shadows, **2**:22*l*
rainy tropics, **6**:21
   *see also* rain forests
ranch farming, of fish, **1**:35
ranching, **2**:14*l*
   deforestation and, **2**:14-15
rangelands, **5**:57-58
   grazing on, **3**:32
   soil conservation in, **5**:110
rangers, **3**:108*l*
rat bite fever, **5**:78*l*
Rathje, William, **3**:5
rats, **5**:78
rat snakes, **5**:126-27
rattlesnakes, **5**:66
raw sewage, **5**:99*l*
reclaiming resources, **2**:63*l*
Reclamation, Bureau of (U.S.), **1**:76-77;
     **5**:59
reclamation, legislation on, **4**:91; **5**:59,
     80, 125
Reclamation Act (U.S., 1902), **5**:59, 80
recombinant DNA research, **3**:13-14
recycling, **3**:6, 108; **5**:59-62; **6**:38
   as alternative to landfills, **5**:116
   of aluminum, **1**:27
   container deposit legislation for,
     **1**:120-21
   of industrial wastes, **3**:82
   labeling for, **3**:104
   of plastics, **5**:10, 23
red algae, **1**:23
red-backed salamanders, **1**:29
red tide, **1**:23
reforestation, **2**:26*l*
regolith, **6**:54
Reimer, Charles, **4**:119
rejection (biological), **3**:14*l*
relative humidity, **6**:53
rems (units of radiation), **5**:47, 47*t*
renewable resources, **4**:59-61; **5**:62-65
   conservation of, **1**:116-17
   forests as, **2**:127-28
   solar energy as, **5**:113

temperate rain forests, **1**:67-68; **5**:53
temperate soil, **5**:109
temperate zones, **5**:35
temperature, **6**:53
　　glaciers and, **3**:24
　　global warming and, **3**:35-36
　　of ocean water, **4**:9, 11, 83-84
　　　thermal energy from, **4**:89-90
　　thermal inversions, **5**:106
　　in troposphere, **6**:22
Tennessee Valley Authority (TVA), **6**:6-7
10% law, **2**:66
terminal moraines, **3**:23
termites, **4**:21*l*, 36
terracing, **5**:111; **6**:7-8
terrestrial habitats, **3**:46-47
terrestrial organisms, **1**:2-3, 2*l*
tertiary treatment
　　of sewage, **5**:100
　　of wastewater, **6**:41
Texas
　　Aransas National Wildlife Refuge in,
　　　**4**:53
　　snakes and lizards in, **5**:66-67
theories, **1**:121*l*
thermal energy, **2**:134
　　cogeneration of electricity with, **1**:109
　　ocean thermal energy conversion,
　　　**4**:89-90
　　thermal inversions, **5**:106
thermal water pollution, **3**:18; **5**:31;
　　**6**:8, 47
thermodynamics, laws of, **6**:8-10
Thoreau, Henry David, **6**:10
threatened species, **6**:66
Three Mile Island Nuclear Power plant
　　(Pennsylvania), **4**:65-66, 79,
　　**6**:10-11
tidal energy, **2**:134; **6**:11-12
tides, **1**:26*l*
　　intertidal zones and, **3**:92-95
　　ocean currents and, **4**:86
tigers, **6**:12-13
timber logging
　　deforestation and, **2**:14
　　in national forests, **4**:40
　　northern spotted owls and, **4**:73-74
　　tree farming and, **6**:18-19
tissues, **2**:105*l*
toads, **1**:27
tobacco smoke, **5**:102-3
toothed whales, **6**:58
topography, **5**:107
topsoil, **5**:108; **6**:13-14
　　erosion of, **1**:10; **2**:78-81; **5**:110
　　　contour farming to prevent, **1**:122
　　　by overgrazing, **4**:103-4

formed by decomposition, **2**:11
　　of grasslands, **3**:32
*Torrey Canyon* (oil tanker), **4**:95
Tortuguero National Park (Costa Rica),
　　**5**:95
tourism
　　ecotourism, **2**:48, 129
　　in Galápagos Islands, **3**:4
toxicity, **3**:116*l*
toxic metals, **6**:46
Toxics Release Inventory (TRI), **5**:130;
　　**6**:16, 17, 48
toxic substances, **4**:122*l*
　　chemicals
　　　as marine pollution, **4**:11
　　　PCBs, **4**:119-20
　　in landfills, **5**:32
　　legislation on, **1**:19; **2**:77; **6**:14-15
　　ocean dumping of, **4**:89
　　wastes, *see* hazardous wastes; toxic
　　　wastes
　　in water supplies, **5**:82; **6**:47
　　　in private wells, **5**:83
Toxic Substances Control Act (U.S.,
　　1976), **2**:77; **6**:14-15
toxic wastes, **6**:15-17, 16*c*
　　international trade in, **6**:18
　　legislation on, **5**:130
　　seabed disposal of, **5**:91-92
　　from tailings, **6**:4
toxins, **5**:14*l*, 15
　　in food chains and food webs, **5**:38
Toynbee, Arnold, **3**:80
trade winds, **4**:86
tragedy of the commons, **6**:18
traits, **4**:61*l*
　　genetics of, **3**:8
　　natural selection of, **4**:62
transportation
　　automobiles for, **1**:46-47
　　energy efficiency and, **2**:64
　　of hazardous wastes, **3**:55
　　　legislation on, **3**:50-51
　　mass transit, **4**:15-17, 16*t*
Transportation, U.S. Department of, **2**:73
　　Hazardous Materials Transportation
　　　Act enforced by, **3**:50
tree farming, **6**:18-19
treeline, **6**:24*l*
trees
　　acacia trees, **4**:36-37
　　American chestnut tree, **2**:119
　　arboreal habitats in, **3**:46
　　in Arctic, **1**:39
　　coniferous, **6**:2
　　conifers, **1**:113
　　in deciduous forests, **2**:8-10

farming of, **6**:18-19
forestry of, **2**:125-26
in forests, **2**:119-22
global warming and, **3**:27
killed by gypsy moths, **3**:43
mangroves, **4**:7-8
pine trees, **2**:104
pollination of, **5**:29
products of, **2**:123-25
in rain forests, **5**:54
riparian, **5**:73
in savannas, **5**:90
silviculture and, **5**:104-5
softwood, **5**:106-7
Spanish moss on, **5**:138*l*
in Yosemite National Park, **6**:75
*see also* forests
Triassic mass extinction, **4**:14
tributaries, **5**:76*l*
trickling-filter method, **6**:41
trihalomethanes (THMs), **6**:48
tritium, **5**:52
trophic levels, **1**:65, 120; **2**:42; **6**:19-21
　　energy flows in, **2**:46
　　in energy pyramids, **2**:65-66
　　energy transferred between, **6**:10
　　in food webs, **2**:117
tropical deciduous forests, **2**:121
tropical rain forests, *see* rain forests
tropical seasonal forests, **1**:68
tropical soil, **5**:109
Tropic of Cancer, **6**:21
Tropic of Capricorn, **6**:21
tropics, **6**:21-22
tropopause, **6**:22
troposphere, **1**:44; **6**:22-23
　　weather in, **6**:52
tuatara (reptile), **5**:66
tularemia, **5**:78*l*
tumors, **1**:80, 81
tuna, **6**:23
　　dolphins and, **2**:34-35, 71
　　mercury in, **4**:19
tundras, **1**:41, 66; **6**:24-25
　　as habitats, **3**:47
　　permafrost of, **4**:21*l*
turbines, **4**:89-90, 89*l*
turbogenerators, **3**:16, 17*l*
Turco, Richard P., **4**:81
turtles, sea turtles, **5**:67, 95-96
typhus fever, **2**:5*l*

# ◗ U

ultraviolet light, **2**:55; **6**:26
ultraviolet (UV) radiation, **2**:55; **6**:26-27, 26*c*

absorbed by ozone layer, **4**:108, 111
as carcinogen, **1**:86
DNA damaged by, **2**:31
energy in, **6**:26*g*
impact on CFCs of, **1**:91
in oxygen cycle, **4**:108
phytoplankton endangered by, **5**:15
in stratosphere, **5**:124
understory layer (in forests), **2**:9*l*, 120;
   **5**:53
Union Carbide (firm), **1**:53-54
United Nations
   Control of Transboundary Movements
      of Hazardous Wastes and Their
      Disposal, treaty on, **6**:17
   Environmental Programme of, **6**:28-29
   International Atomic Energy Agency
      of, **3**:89-90
   International Registry of Potentially
      Toxic Substances of, **6**:18
   Law of the Sea Convention and,
      **3**:114-15; **4**:89
   World Bank and, **6**:72
United Nations Conference on the
     Human Environment (1972,
     Sweden), **2**:68; **6**:27, 29
United Nations Earth Summit (Brazil,
     1992), **6**:27-28
United Nations Environmental
     Programme (UNEP), **3**:36, **6**:27-29
United Nations Sofia Agreement (1989),
     on acid rain, **1**:6, 19
United Nations Treaty on the
     Non-Proliferation of Nuclear
     Weapons, **3**:90
United States
   Agriculture, Department of, **2**:17-19
     food pyramid of, **3**:60*c*
     Forest Service of, **2**:126-28; **4**:39-40,
      **5**:15-16
     Soil Conservation Service of, **5**:110
     soils classified by, **5**:109
     sustainable agriculture program of,
      **4**:100
   Army Corps of Engineers, **1**:41-42
   Atomic Energy Commission, **1**:45
   bald eagle as symbol of, **1**:50
   Commerce, Department of
     National Oceanic and Atmospheric
      Administration of, **4**:41-42
     National Ocean Service of, **4**:12
     National Weather Service of,
      **4**:48-51; **6**:54
   Council on Environmental Quality,
     **1**:129-30; **4**:38
     Global 2000 Report of, **3**:25
   DDT banned in, **1**:90; **2**:6-7

Dust Bowl in, **2**:37-38
endangered species in, **2**:61
Energy, Department of
   Federal Energy Regulatory
     Commission in, **2**:100-101
   Hanford nuclear facility of, **3**:49
environmental education in, **2**:67-68
Environmental Protection Agency,
   **2**:76-77
   drinking water regulated by, **6**:43
   environmental justice studies of,
     **2**:75
   Office of Waste Minimization of,
     **6**:17
   Superfund under, **5**:129-32
Food and Drug Administration, **3**:60
governmental organizations with
     environmental responsibility,
     **6**:95-96
Great Lakes of, **3**:33-34
history of environmental events and
     legislation in, **6**:79-83
indigenous people of, **3**:78
infant mortality in, **3**:83
Interior, Department of, **2**:19-20
   Bureau of Land Management of,
     **1**:76
   Bureau of Reclamation of, **1**:76-77;
     **5**:59
   Fish and Wildlife Service of, **2**:61,
     62, 106-7; **6**:56, 69
   national grasslands under, **4**:40-41
   National Park Service of, **4**:43-47;
     **5**:43, **6**:66
   Office of Surface Mining,
     Reclamation, and Enforcement
     of, **4**:91
   wildlife refuges of, **4**:51-54
land use in, **3**:110*t*
land utilization in, **5**:110*c*
Law of the Sea Convention and,
   **3**:115
mass transit in, **4**:15-17, 16*t*
National Conservation Commission,
   **5**:80
Occupational Safety and Health
   Administration, **4**:71
Office of Technology Assessment,
   **3**:80
oil production in, **5**:5
old-growth forests in, **4**:97
PCBs banned in, **4**:119
pesticide use in, **5**:2
public lands of, **4**:34-36; **5**:42-45
   national forests, **4**:38-40
   national parks, **4**:42-46; **6**:66
   national seashores, **4**:48

national wildlife refuges, **4**:51-54
   Reclamation Act and, **5**:59
   Tennessee Valley Authority, **6**:6-7
   zoos in, **6**:77
   *see also* legislation
U.S. Synthetic Fuels Corporation, **5**:139
upwellings, **4**:86*l*
uranium, **1**:75; **6**:29
   in nuclear fission, **4**:75-76
   radioactive wastes from, **5**:48-50
   radioactivity of, **5**:51
   radon and, **5**:52
urogenital conditions, **3**:84*l*, 88
Utah, Bingham Canyon copper mine in,
   **4**:99

# V

Van Helmont, J.B., **5**:11
variations
   in evolution, **2**:90
   genetic diversity and, **3**:12
vascular plants, **2**:102; **5**:20-21
vegetation
   eutrophication of, **2**:85
   *see also* plants
Venice (Italy), **1**:37, 38
ventilation systems, **5**:103
Vermont, Green Mountain National
     Forest in, **4**:40
vertebrates, **3**:96; **6**:30-33
   birds as, **1**:72
   fish as, **2**:105
   mammals as, **4**:4-6
   reptiles, **5**:65-67
Vietnam War
   defoliants used during, **2**:12-13
     in mangrove swamps, **4**:8
   herbicides used during, **3**:63
viruses, **6**:34
   as carcinogens, **1**:86
   used in genetic engineering, **3**:13
visible wavelengths, **2**:54-55
vitamins, **3**:58-59, 59*t*
   deficiency diseases, **4**:3, 3*t*
volcanism, **6**:34-35
volcanoes, **5**:40-41; **6**:35
   eruptions of, **4**:55-56
   plate tectonics and, **5**:25
vultures
   California condors, **1**:78-80
   as scavengers, **1**:87

National Weather Service and, **4**:42, 48-51
natural disasters, **4**:56
precipitation, **5**:37
thermal inversions, **5**:106
in tropics, **6**:21
in troposphere, **1**:44
weathering, **6**:54
soil formed by, **5**:107
weather satellites, **4**:48, 50*l*
Wegener, Alfred, **1**:121-22; **5**:23
wellheads, **5**:82, 83*l*
wells, **1**:35*l*, 37
artesian, **1**:42
injection wells, **6**:16*l*, 38
private, **5**:83
Wen Wang, **6**:77
West Virginia, Buffalo Creek dam in, **5**:96
wetlands, **1**:59; **6**:54-56
estuaries and, **2**:83
mangrove swamps, **4**:7-8
methane released by, **3**:26
pollution of, **2**:84-85
salt marshes, **5**:86-87
swamps, **5**:30, 31*l*
in wildlife refuges, **4**:53
Wetlands Protection Act (U.S., 1986), **6**:56
whales, **1**:32; **6**:56-59, 67
dolphins as, **2**:32
environmental ethics and, **2**:70
International Convention for the Regulation of Whaling, **3**:90-91
International Whaling Commission and, **3**:92
legislation protecting, **4**:9
whaling, **3**:90-92; **6**:58
wheat, **3**:38
White House Office on Environmental Policy, **1**:129; **4**:38
whooping cranes, **1**:74; **4**:53; **6**:67
wilderness, **6**:60-61
conservation easements for, **1**:118
National Wilderness Preservation System for, **5**:42-43
old-growth forests as, **4**:97
Pinchot on, **5**:16
Yosemite National Park as, **6**:75

Wilderness Act (U.S., 1964), **3**:37; **5**:42; **6**:60-62
Wilderness Society, **6**:63
wildlife, **6**:63-65
acid rain's effects on, **1**:4-5
in Arctic, **1**:38-39
Arctic National Wildlife Refuge for, **1**:40-41
bacterial diseases of, **1**:50
conservation of, **6**:65-68
demographics of, **2**:17
effects of oil pollution on, **4**:93
effects of pesticides on, **4**:122-23
Fish and Wildlife Service and, **2**:106-7
hunting of, **2**:59
management of, **3**:118; **6**:68-69
in national forests and grasslands, **2**:127
national wildlife refuges for, **4**:51-54
rehabilitation of, **6**:69-70
smuggling and poaching of, **1**:123-24
wildlife habitats, **1**:118
wildlife management, **3**:118; **6**:68-69
Wild and Scenic Rivers Act (U.S., 1968), **2**:3; **6**:59-60
Wilson, Edward Osborne, **1**:57-58, 68-69; **3**:47, 48; **4**:2, 59; **6**:64, 70
windmills, **2**:133-34; **6**:70-72
wind power, **1**:26; **2**:52, 133-34; **6**:70-72
winds, **6**:53
as energy source, **6**:70-72
erosion caused by, **2**:79; **5**:108-9
ocean currents and, **4**:86
pollination by, **5**:29
wolves, **4**:6; **5**:38
women
ecofeminism and, **2**:40
in fertility rates, **2**:103
wood, **2**:124
as biomass energy, **2**:133
as fuel, **2**:134
Global 2000 Report on, **3**:25
harvested from national forests, **2**:127
softwood, **5**:106-7
woodpeckers, ivory-billed, **3**:100
World Bank (International Bank for Reconstruction and Development), **6**:72
World Wildlife Fund, **2**:8

Wyoming, Yellowstone National Park in, **1**:vii, **3**:112; **4**:34, 43, 46; **5**:42; **6**:74
fires in, **2**:104-5; **4**:56-57; **5**:129
grizzly bears in, **3**:40
as wildlife refuge, **6**:66

# X

xenobiotics, **6**:73
x rays, **2**:55; **6**:73
xylem (in plants), **5**:20-21, 107

# Y

yeasts, **2**:135*l*, 138
yellow algae, **5**:17
yellow fever, **2**:5*l*
Yellowstone National Park (Wyoming), **1**:vii, **3**:112; **4**:34, 43, 46; **5**:42; **6**:74
fires in, **2**:104-5; **4**:56-57
secondary succession following, **5**:129
grizzly bears in, **3**:40
as wildlife refuge, **6**:66
Yosemite National Park (California), **4**:43; **6**:75
Muir and, **4**:33-34
Pinchot and, **5**:16
Yucca Mountain (Nevada), **5**:50

# Z

zebra mussels, **2**:59, 91; **3**:34
Zeidler, Othmar, **2**:5
zero population growth, **6**:76
zone of aeration, **6**:51
zone of saturation, **1**:36; **3**:83; **6**:51, 76
zooplankton, **5**:14, 17; **6**:78
zoos, **6**:77-78
gene banks of, **3**:10
mountain gorillas captured for, **2**:128, 129
zooxanthellae (algae), **1**:22
zygomycetes, **2**:137